CAT

COLOSSAL CATERPILLAR

THE ULTIMATE EARTHMOVER

ERIC C. ORLEMANN

MBI Publishing Company

CATERPILLAR
797
797
202

First published in 2002 by MBI Publishing Company, Galtier Plaza, Suite 200, 380 Jackson Street, St. Paul, MN 55101-3885 USA

MBI Publishing Company books are also available at discounts in bulk quantity for industrial or sales-promotional use. For details write to Special Sales Manager at Motorbooks International Wholesalers & Distributors, Galtier Plaza, Suite 200, 380 Jackson Street, St. Paul, MN 55101-3885 USA.

Library of Congress Cataloging-in-Publication Data Available

ISBN: 0-7603-0874-8

Edited by Kris Palmer
Designed by Dan Perry

On the front cover:
With its nominal 360-plus-ton capacity rating and 3,211 flywheel horsepower V-24 diesel engine, the incredible Caterpillar 797 is the world's largest and most powerful mechanical drivetrain mining hauler available to the mining industry today. *ECO*

On the frontispiece:
A worker applies the final logo decals on a D11R at Caterpillar's East Peoria Building SS Assembly Plant. Recognized the world over as the largest producer of heavy equipment, Caterpillar builds some of the largest, most technologically advanced and productive machines ever offered to the mining industry. *ECO*

On the title page:
A Caterpillar 797, equipped with a special MSD coal body, slowly makes its way out of the pit, at a mine site in the Powder River Basin coal mining area of Wyoming. With a massive load of 385 tons of coal on its back, a trail of hot exhaust is left in the air from the trucks V-24 diesel. Soon it will be back for another load and the process will begin again. *ECO*

On the back cover: (large)
The early Caterpillar 994 loaders were large and powerful machines. With its 1,250 flywheel horsepower, V-16 Cat 3516 diesel, and standard 23-cubic-yard bucket, the 994 was one of the largest mechanical drive front-end loaders available to the mining industry. Because of early tire wear problems, many customers installed tire chains on their 994 loaders to protect them from rock cuts and abrasions. *ECO*

(small top) It is late October 2001, as the first 5230B goes through its final engine tests at the Decatur assembly plant. In a couple of weeks, it will be ready for its world unveiling. In shipping, the 5230B breaks down into individual modular sections for easy transport. The main superstructure, cab, and swing circle all ship from the Decatur facilities. The crawler assemblies, bucket, and main boom all ship to the customers assembly site from other Cat manufacturing plants. *ECO*

(small bottom) Once all of the main powertrain components check out, the 24H will make its way outside of the main plant for preliminary running and braking tests. The graders 24-foot moldboard will not be installed until the machine reaches its final assembly destination. *ECO*

Printed in Hong Kong

CONTENTS

ACKNOWLEDGMENTS

It seems like only yesterday that Caterpillar Corporate green-lighted this project. In fact, it was December 1999. My, how time flies when you're having fun! In reality, I was producing *Colossal Caterpillar* while working on three other book assignments. Needless to say, it has been a very busy last two years.

I have the fantastic people of Caterpillar, Inc., to thank for making this book a reality. In the front offices and the manufacturing plants, I have had the utmost cooperation and guidance from the employees who actually design, build, market, test, and ship these giants of the mining industry. Though I have come in contact with numerous Caterpillar personnel over the years, a few individuals bear special mention for the time and effort they have given me to ensure that the mining machines of the company were covered in every way possible without revealing the manufacturing trade secrets that make these massive pieces of equipment possible. I would like to personally express my sincerest thanks to Jeff Hawkinson, Sharon L. Holling, Pete J. Holman, John H. Ingle, Dave Faber, Michael D. Ray, Phil Holthaus, Bernie Nolte, Carl M. Volz, Angela H. Myers, Eleanor Alsbury, Ron Nusbaum, Eldon D. Oestmann, Bob Lapkee, Benjamin S. Cordani, Bill Roberts, Ed Staley, Dan Winslow, William F. Pullman, Catherine Wells, Fred Dully, Kent H. Clifton, Ron Kuykendall, Stephan Ihnken, James E. Gee (retired), and Larry Clancy (retired).

I would also like to thank the following mining companies and personnel who have allowed me to capture some of Caterpillar's latest mining giants at work, doing the jobs they were built to perform. They are: Christine L. Taylor, Beth C. Sutton, Bob Heimann, and Greg Dundas of the Peabody Group and the Powder River Coal Company; Thomas J. Lien and William M. Dalton of RAG Coal West, Inc.; Nance Dania of AEP Central Ohio Coal Company; and Greg Halinda of Syncrude Canada Ltd.

Last but not least, I thank my good friends Keith Haddock and Urs Peyer for their additional research and photographic assistance.

Eric C. Orlemann
Decatur, Illinois

INTRODUCTION

Caterpillar, Inc., is recognized the world over as the leader and number one producer of heavy equipment. Its logo and company abbreviation "CAT" are the trademarks that marketing and advertising people dream about. Just hearing the company's name conjures up images of ground-pounding, thundering machines of dynamic yellow iron, pushing, loading, and hauling tons of earth. It's hard to travel far and not see the famous "yellow iron" working along the roadways, at a home or business construction site, or in the far-off distance on some type of earthmoving job whose outcome we can only imagine. Caterpillar's machines, and the work they have accomplished, are everywhere in our daily lives.

But there is a class of Caterpillar creations rarely seen by the public. These are the large mining and quarry equipment lines. These colossal Caterpillars work out of view for most of their lives. Some work in the confines of large stone and rock quarries, while others toil in remote mining operations around the world. These machines are designed and built to dig, load, and haul thousands of tons of material daily, 24 hours a day, 7 days a week, 365 days a year if need be. This is especially true for the ultralarge mining equipment lines, which are expected to work shift after shift, resting only for personnel changes and scheduled maintenance. In this type of environment, only the strongest, best-designed machine will do. When equipment of this magnitude is out of service for any length of time, the costs in lost production to the owner or operator can be staggering.

Over the decades, Caterpillar has been viewed as the undisputed leader in track-type tractor development and sales, and rightly so. But there is more to the company than its highly productive dozers. Various other tracked and wheeled product lines have been introduced over the years to further strengthen the company's position in the heavy equipment marketplace. Motor graders were added in the 1930s, towed scrapers in the 1940s, self-propelled scrapers, wheel and track loaders in the 1950s, off-highway trucks in the 1960s, hydraulic excavators in the 1970s, and rubber-tracked agricultural tractors in the 1980s. To finish out the twentieth century and enter the twenty-first, the 1990s would be the decade of the company's new mining equipment giants.

Caterpillar products have always been an important ingredient in the world mining market. But only recently has the company challenged and surpassed manufacturers of the biggest pieces of mining machinery. Companies such as Euclid, Terex, WABCO, and Unit Rig dominated the large mining hauler industry with diesel-electric drivetrain designs. Throughout the 1960s and 1970s, Caterpillar only offered three sizes of trucks, with the largest, the 777, rated at an 85-ton capacity. On the other hand, the builders of the large diesel-electric-drive trucks were supplying the industry with 150- to 235-ton-capacity haulers, and even one 350-ton brute. This story played out with the other equipment lines as well. Smaller manufacturers were filling the niche markets, ignored by Caterpillar because of their perceived low profitability as it related to equipment sales. But as world mining operations continued to expand, so did the industry's appetite for the largest mining machines.

To address the needs of its mining customers, as well as its dealers, Caterpillar established the Mining Vehicle Center, better known as the MVC, at its Decatur, Illinois, facilities. Planning on the MVC started in 1987, with full operations commencing in 1988. The MVC was responsible for developing a new generation of mining tools far larger than anything the company had ever built. These early projects included a wheel loader (Model 994), an off-highway hauler (793), a motor grader (24H), and a hydraulic excavator (5130). More equipment designs would follow as each machine program was advanced. By the mid-1990s, the Decatur assembly plant was renamed the Mining & Construction Equipment Division. Though the name was changed, Decatur was still responsible for the design and marketing of most of the company's specialized mining equipment.

On January 1, 2001, a new group within the company was established to oversee all marketing and planning of the large mining equipment. Referred to as the Caterpillar Global Mining Division, it makes sure that each product group is building the right type of tools to serve the needs of the mining industry. The Global Mining Division is located at Caterpillar's corporate offices in Peoria, Illinois.

Caterpillar's mining equipment designs are assembled at many of its various manufacturing plants. Though the company has over 90 plants in 19 countries worldwide, all of the large mining machines are built in Illinois. The East Peoria assembly plant is responsible for the massive D11R. The Aurora facility produces the 992G and 994D wheel loaders, along with the 854G wheel dozer. The 24H grader, the 5110B, 5130B, and 5230B hydraulic excavators, and all of the off-highway haulers, including the monstrous 797, ship from Decatur. It has been said that Peoria, Illinois, is the earthmoving capital of the world. The state of Illinois can also rightly be referred to as the mining equipment capital of the world.

Colossal Caterpillar takes an in-depth look at the massive mining machines of Caterpillar, Inc. All are covered in great detail, from the giant 797 hauler, to the legendary D11R track-type tractor, giving the reader a look at how these giant Cats are built, and how they work. The emphasis of this book is the large mining vehicles. It is not intended to tell the entire history of Caterpillar, nor highlight all of the various types of fantastic machine models produced over the company's long and distinguished history. This book celebrates the biggest, baddest, heaviest, most powerful, and most awe-inspiring chunks of yellow iron ever to haul, load, push, grade, and dump massive piles of earth. They are the colossal Caterpillars.

In early 1996, the new D11R took over for the D11N. Though initially rated at 770 flywheel horsepower, the D11R received a substantial increase in grunt in 1997, as power was increased to 850 flywheel horsepower. The unit pictured is shown equipped with the optional black antiglare paint job and upper track carrier rollers. *ECO*

CHAPTER ONE

The Living Legend
D11R BULLDOZER

For most, the mention of the word "Caterpillar" conjures up an image of a large yellow track-type crawler tractor bulldozer. With an entire company founded on such a machine, it is no wonder that just about everyone in the world associates Caterpillar with this type of earthmover. Though the company offers a diversified line of products and services to meet the earthmoving, forestry, material handling, agricultural, and power generating industries, it is the track-type tractor offerings that are the company's heart and soul. And nowhere is this more evident than with its top-of-the-line crawler tractor model—the massive D11R.

The D11R represents the largest and most powerful track-type tractor ever to roll out of the company's East Peoria, Illinois, tractor assembly plant. This giant bulldozer contains the hallmarks that have made Caterpillar the number one producer of heavy equipment worldwide. The D11R's innovative design and technology, and thus its power, serviceability, availability, and productivity, set it apart.

It's hard to believe, but Caterpillar's elevated drive sprocket system of today owes its very existence to this little single-seat contraption. Built in 1965 out of mainly discarded lawnmower components, it proved the design concept of the "triangle-track drive" offered superior tractive capabilities. *ECO Collection*

To understand how the D11R series was designed and developed, you first have to look back to the year 1977. That year Caterpillar introduced its first production elevated-drive sprocket track-type tractor design, the legendary D10. The D10 not only was the company's first "hi-drive" offering, it was also the largest single-engined crawler tractor Caterpillar had yet designed.

The company developed the D10 and the elevated-drive sprocket concurrently over a 13-year period starting in 1965. It was in that year that a Caterpillar research engineer by the name of Bob Purcell started to investigate other forms of track drive arrangements that might offer tractive capabilities superior to the designs currently in favor at the time. He found that a triangle-track drive arrangement, or elevated sprocket, in conjunction with an oscillating lower track bogey, seemed to offer the most promise. To test his design theories, he had the Caterpillar model shop fabricate a triangle-track assembly with oscillating bogies. This was then incorporated into an experimental test tractor built out of odds and ends from a riding lawnmower. The simple tractor steered by means of an articulated frame, with a single track assembly in the front and two wheels on the rear. This experiment demonstrated how the oscillating bogie kept the track firmly on the ground when driven by an elevated sprocket drive. Tested at Caterpillar's Peoria Proving Ground (PPG), it performed far beyond anyone's expectations. To further prove the triangle-track's tractive capabilities, testing personnel hitched up the experimental rig to a Jeep. They were amazed once again as the little test tractor pulled the Jeep around the PPG without breaking traction. The engineers who witnessed the demonstrations knew that this was a significant design innovation, but more testing would have to be done to prove to management that this was the wave of the future. Purcell's little invention would change the way Caterpillar track-type crawlers would be designed and built in the not too distant future.

Following months of detailed soil-bin testing analysis, management approved moving the triangle-track drive system to the full-scale experimental

The engineering and design team responsible for the D10 track-type tractor project gathered on April 2, 1977, with the first pilot preproduction unit. Eldon Oestmann is the person holding the sign; to the left of him, front row, is Jim Duke; and to the right are Ron Krolak and George Alexander. These four engineers are named as the inventors on Caterpillar's first elevated-sprocket patent for use on a track-type tractor. *Caterpillar, Inc.*

testing stage. Yet it was not to be a crawler tractor that was to benefit from this concept drive system. Instead the system was fitted to the 988 wheel loader. Caterpillar Research saw the triangle-track assemblies as a way for the wheel loader's tires to avoid punctures from sharp rocks. Built in April 1967, the 988 Track Loader showed promise, but the complexity of the track assemblies, and their added cost, proved a liability. In October 1968, engineers tried the 988 with track assemblies mounted to the front axle only, with standard wheels and tires in the rear, in the hopes of controlling costs. But it fared no better than the first prototype design.

In 1969, as Caterpillar Research was tinkering around with triangle-track-drive loaders, another Caterpillar engineering team was starting to assess possible drive layouts for a new, ultralarge bulldozer to be called the D10. Because of the steady increase in the scale of construction and mining operations the world over, Caterpillar management felt that a crawler dozer larger than the company's top-selling D9 series would be needed in the not too distant future. An engineering team was set up to design a new concept dozer that would fill the company's future needs. To assist this team, a group of engineers was also invited over from Cat Research.

Over the months, the group of engineers came up with three possible directions for the D10 project: One design utilized a quad-track arrangement with an articulated chassis. Another was a skid-steer quad-track layout. But the designers ultimately felt that these designs would produce excessive ground pressure. The group then decided to take the elevated sprocket design of the triangle-track drive and incorporate it into a two-track dozer arrangement. By November 1969, this concept emerged as the engineers' choice to proceed to the next level in the design process—a full-size mock-up.

The group constructed a full-size mock-up of the D10 concept from wood and cardboard to better illustrate the overall size and design characteristics to upper Caterpillar management. Management liked what it saw and authorized the

In 1970, Caterpillar engineers converted a standard D9G track-type tractor into a test bed for the elevated-sprocket drive program. Completed in June of that year, it provided a wealth of information and data that would eventually lead to the production of the D10. It is shown here, undergoing an undercarriage demonstration test in 1972. *ECO Collection*

building of a test bed tractor. In February 1970, the team started converting a standard D9G dozer into an experimental elevated-drive sprocket test-mule tractor. This endeavor was completed in June of that year. This modified D9G utilized an elevated sprocket system with a resilient mounted bogey undercarriage. The resilient undercarriage allowed the track rollers to float over obstacles for improved dozer and operator ride, and better traction. The system also reduced impact loading on the rollers and roller frames. In 1973, after two full years of collecting engineering data from the D9G test mule, it was time for the company to begin fabricating a full-size D10 prototype in the iron.

Early in the D10 project, the company set five primary goals for the tractor: high productivity, modular design, simplified maintenance, high operator efficiency, and transportability. Those in upper management who had stuck their necks out for the D10 knew those goals had to be met, because if the customer was unwilling to accept the elevated sprocket system, or if it did not work as promised, many of their careers would be dozed under, most likely taking a few engineers with them. Project supporters were not only risking millions of dollars in developmental costs, they were also playing around with the very product line that formed the essence of the company's public image. A misstep with the track-type tractor line would have far-reaching implications for the company, both financially and psychologically.

On July 13, 1973, the first D10 prototype, identified as the X1, was completed. The following week, it was presented for the first time to company officers and plant managers at their summer corporate meeting, held at the Caterpillar Tech Center in Mossville, Illinois. This prototype featured an open Rollover Protective Structure

(ROPS), and two upper track carriers. After the meeting, the D10X1 was secretly shipped over to the PPG for further evaluations. Six weeks later, the second prototype, D10X2, was ready for testing. D10X1 would spend its entire testing life at the PPG, while D10X2 was assigned to the Caterpillar Arizona Proving Ground. The second prototype resembled the first, except it was equipped with a full ROPS cab with air conditioning to cope with the Arizona heat.

Engineers used information gathered over two years of evaluation testing of the first two D10 prototypes in constructing the next two test tractors in 1975, D10X3 and X4. Though these prototypes resembled the first two units, both featured fully enclosed ROPS cabs, and improved resilient undercarriages, now with the elimination of the two upper track carriers on each side. D10X4 looked much like X3, but featured a more refined ROPS cab design.

During the testing phase of the X3 and X4, Caterpillar started building an entirely new tractor assembly plant in East Peoria in 1976. This plant, identified as Building SS, would be the future home of the D10, as well as other elevated-sprocket design tractors that were soon to be. In late March 1977, the first of 10 pilot D10 dozers (D10P1) came off of a temporary assembly line

In September 1977, Caterpillar officially announced the introduction of the D10 at a special event held at its demonstration area. The refurbished experimental prototype D10X2 dozer with its one-of-a-kind "riveted" ROPS cab was on hand for the ceremony. *Caterpillar, Inc.*

This D10 unit, one of the original 10 pilot preproduction machines, was at work in May 1977 at Caterpillar's Arizona Proving Ground. Note the single exhaust stack and preliminary ROPS cab design. This ROPS configuration would change slightly in the final production tractors starting in 1978. *Caterpillar, Inc.*

set up in Building SS, still under construction. And not a moment too soon—while the D10 was coming to fruition, industry competitors were starting to put a horsepower squeeze on the company's front line big dozer, the D9H.

The D9 series was Caterpillar's premier large dozer. It was the company's most powerful single-engined tractor model and was extremely reliable when maintained properly. But it was not the most powerful in the industry. Crawler tractors such as the Euclid/Terex 82-80 DA (TC-12) and the Allis-Chalmers HD-41 had power outputs exceeding the Cat D9G and D9H model lines. The D9G, with 385 flywheel horsepower, and the D9H, with 410, were no pushovers, but the Euclid/Terex 82-80 DA could lay claim to 440 flywheel horsepower. And the big Allis-Chalmers? In 1970, after years of prototype testing, the Allis Chalmers HD-41 was the first commercially available crawler dozer to break the 500-horsepower barrier, with 524 flywheel horsepower on tap. In 1974, this model became the Fiat-Allis 41-B with the same power output. Though both of these competing designs offered big power advantages over comparable D9G/H dozers, their reliability records were another matter entirely. The HD-41/41-B was prone to electrical fires and final drive failures, while the Euclid/Terex offering, with its twin engine configuration, had frequent problems in the field. Of more concern to Caterpillar was Komatsu's increasing share of the North American market. The company's concerns rose further in 1975 when Komatsu introduced what it called the world's first superdozer, the D455A.

In 1986, Caterpillar replaced its legendary D10 track-type tractor with an even more powerful machine—the D11N. The D11N was equipped with a Cat 3508 V-8 diesel engine, which replaced the previous model's D348 V-12 unit. The 3508 was capable of producing 770 flywheel horsepower, 70 more than the D348. Pictured is the prototype D11N undergoing testing in September 1985. Note the Cat logo D11N nomenclature. This would be redesigned by the time full-production units started coming off the assembly line in February 1986. *Caterpillar, Inc.*

With a flywheel horsepower rating of 620, and a 42-cubic-yard bulldozing blade, the term super-dozer wasn't far off the mark. And Caterpillar still was two years away from introducing the D10.

To help keep potential customers from turning to these more powerful offerings, Caterpillar introduced twin tractor configurations in the form of the SxS D9G/H, and DD9G/H. The side-by-side SxS D9G model was introduced in 1969 with a power rating of 770 flywheel horse-power, and was marketed as a high-production land reclamation dozer. In 1974, power output increased to 820 flywheel horsepower with the introduction of the SxS D9H. The DD9G, intro-duced in 1968, was an outgrowth of the Quad-Trac D9G originally developed by Buster Peterson of Peterson Tractor Company in late 1963. This configuration had two D9G tractors attached one behind the other, and was marketed as a powerful pusher for scraper loading. In 1974, it became the DD9H, with power increases the same as the side-by-side configurations.

Caterpillar would eventually produce 10 pilot D10 dozers. All were put into the field at various

CATERPILLAR
6278

specially selected job sites across the United States, including the company's proving grounds. These tractors were almost to production status, except for the final design of the ROPS, which still needed some work.

In September 1977, at Caterpillar's Demonstration Area, the company finally let the big Cat out of the bag. On hand was one of the early experimental D10 units (X2), updated mechanically to pilot machine specifications, except for the ROPS cab. Caterpillar invited its dealers and best customers to the event so they could be the first to see the new dozer up close. The company also used the occasion to announce officially the availability of the new tractor to the heavy construction and mining industries. This announcement was a bit premature, as company engineers were still sorting out some design areas. But by the spring of 1978, full-production D10 units were coming off the assembly line at Building SS. Earthmoving history was being made.

The D10 was a technical and mechanical marvel for Caterpillar. The company had brought all of its engineering resources to bear on this groundbreaking elevated sprocket drive dozer. Along with revolutionary mechanical design features, the D10 also produced some eye-opening industry world record specifications for its day. The D10 produced 700 flywheel horsepower and wielded a U-dozer blade that was 19 feet, 10 inches across and 7 feet high. In full operating trim, the big Cat would weigh in at 191,100 pounds. Its closest rival at the time, the Komatsu D455A-1, was rated at 620 flywheel horsepower and tipped the scales at 178,700 pounds. Advantage Caterpillar. To be fair, the big Komatsu's full U-blade capacity was 41.7 cubic yards, compared to the D10's 35 cubic yards. Though the D455A-1 blade capacity was greater than the D10's, the big Cat was far more productive because of its advanced undercarriage design. The big Komatsu was a fine dozer, one of the best in the world in fact. But the D10 was superior in power, weight, productivity, and most importantly, reliability.

The D10 offered performance like no other dozer in the world. It also had a look that was totally unique. Most notable was the massive elevated sprocket drive system, around which the entire project was designed. Though it gave the dozer the appearance of a top-heavy machine with a high center of gravity, nothing was farther from the truth. With a track contact length of 12 feet, 8 inches, and an overall track width of 12 feet, stability was never an issue in the field. The elevated

The D10 was not only a great dozer, it was also an incredible ripping machine. With its giant ripper shank, it could penetrate 71 inches into the ground and was capable of shattering solid rock. This unit, shown testing in August 1980, is one of the later models equipped with the revised dual exhaust system. Operating weight of this vintage D10, with this equipment, was listed at 190,117 pounds. *Caterpillar, Inc.*

Starting in 1987, Caterpillar Custom Products started to offer its giant hydraulic Impact Ripper option for use on the D11N. In operation, a hydraulic impactor transmitted powerful energy pulses through a specially designed forged-alloy steel ripper shank. It concentrated 450,000 pounds of impact at the ripper tip at 540 times a minute. The resulting shock forces fractured rock not only at the tip, but ahead of it as well. When equipped with this option, the weight of the D11N soared up to 225,950 pounds in full operating trim. *Caterpillar, Inc.*

final drive and sprocket design removed shock and implement loads that caused gear and bearing misalignment. They were also in a less hazardous location, away from water, mud, and rock. This reduced abrasive wear and damage to seals, and promoted a longer productive life for all components concerned. Another bonus to this design was that the final drive modules could be removed individually in the field. By breaking the track, the complete final drive system, or the entire final drive, steering clutch, and brake system could be removed as a unit.

The D10's resilient undercarriage gave the giant's operator a ride quality that was previously unknown in the industry. This was made possible by the use of bogey-mounted track rollers, a feature unique to the D10 at the time. All bogies, four per side, oscillated on sealed and lubricated cartridge pins to conform to rough terrain, putting more track on the ground for better traction. A bogey undercarriage meant that rollers and idlers were almost always in contact with the track link rails, sharing the load with adjoining rollers. Dome-shaped rubber pads, mounted both on the major bogies and roller frames, controlled the resiliency and the travel of the bogies for a smooth ride and lower impact loading.

The D10 was also the company's first large track-type tractor to be based on a modular design for all of its major drivetrain components. Engine and torque converter, power shift transmission, steering clutch, brakes, and accessory drives could all be thoroughly tested before installation. Modular design also reduced repair time and costs because of easy access to the components. Jobs that took days on other large dozers, now took only hours on the D10.

Giving the D10 the necessary dozing muscle was a twin dual overhead cam, Cat D348 PCTA, four-stroke, twin-turbocharged, V-12 diesel engine, rated at 700 flywheel horsepower at 1,800 rpm. This well-proven engine had already put in thousands of reliable hours of operation in the company's 777 hauler and 992B wheel loader model lines. The engine drove a rear-mounted planetary power shift transmission, with three speeds forward and three speeds reverse. Built-in hydraulic modulation permitted unrestricted speed and direction changes under full load.

Up front, the D10 was available with three different dozing blade options, all Cat designed and built. The 10S was a straight blade that measured 18 feet across. The 10U was a full U-blade with a width of 19 feet, 10 inches. The

CAT
CATERPILLAR
D11N

A Caterpillar worker puts the final torque rating on the main bolts connecting the massive left final drive assembly of a D11R to its chassis, as it makes its way down the assembly line in Building SS, in East Peoria, Illinois. The outer hub of the final drive assembly houses the double planetary reduction gear sets. *ECO*

third style was the 10C cushion blade. This special purpose blade, measuring 12 feet, 6 inches wide, had a reinforced cushion plate with a blade recoil mechanism, to absorb impact loads when push loading large scrapers from behind. As mentioned before, the U-blade had the largest capacity, at 35 cubic yards.

At the rear of the dozer was its massive hydraulic ripper. Available in single and multishank configurations, the ripper was designed to break up rock, glacial till, and coal for easier dozing and less blasting—critical when working near populated areas. Up to three multishank arrangements could be used in a variety of general working conditions. The large single-shank ripper was designed for extremely tough materials and deep ripping. The design of the ripper also allowed for the easy removal of the transmission module, since it did not have to be removed from the tractor first. Weight of the D10 with the heavyduty single-shank ripper was 191,100 pounds. With multishank ripper, the unit weighed 190,300 pounds with one shank, and 193,000 pounds with all three in place. If the customer did not need a ripper, the factory installed a rear-mounted 6,000 pound counterweight to offset the massive weight of the front-mounted bulldozing blade.

The D10's modular design also greatly simplified the tear-down process for transporting the tractor. The D10 could travel on two high-capacity rail flatcars, with the main tractor assembly on one, and the dozing blade on another. For over-the-road transport, the dozer could be easily broken down into approximately four truckloads. When shipped in this configuration, the D10 could be assembled or disassembled in two shifts with only four men with the aid of a 12-ton-capacity crane or loader.

Once the main chassis and undercarriage of the D11R come together, it is rolled onto a set of tracks. The tracks are then joined together individually. Here, the final bolts of the right track connecting link are being torqued to their proper specified rating. *ECO*

The D10 hit full production in 1978. Though many longtime customers were skeptical of the elevated sprocket system, once they got their hands on one of the tractors, all reservations vanished. They loved it. It was everything Caterpillar said it would be, and then some. It was a fantastic dozing machine. But it was also equally great as a ripping dozer. In the field, the new dozer was more than 50 percent more productive than the D9H model. The D10 gamble paid off big for Caterpillar. The new giant was simply the most productive track-type tractor ever put into service by the company. The earthmoving industry would never look at dozers in the same way. The elevated sprocket drive system was here to stay.

Initially the company built two D10 model types. The standard tractor, with a 114-inch track gauge, carried a serial prefix code of 84W, while the push-dozer variation, with a 106-inch gauge, carried the 76X prefix. Eventually, the 76X model variation was phased out of production in favor of the standard gauge unit.

Starting in 1980, production D10 tractors were built with twin exhaust pipes instead of a single large unit. This was to accommodate a redesigned exhaust system under the hood. In addition to this change, the twin turbochargers were relocated further away from the firewall, and new air cleaner intakes were installed on top of the hood, behind the exhaust pipes. Before, the air pickups were toward the rear on the sides of the engine cowl. Many units in the field with the old systems were updated during scheduled engine maintenance and rebuild downtimes. Aside from these changes, the D10 went through its working life looking much as it did when first available for sale in 1978. The last D10 came off the Building SS plant's assembly line in January 1986, capping

This D11R is going through a complete set of tests on a running engine to make sure all systems check out as ready for duty. The ductwork piping coming out of the top of the tractor carries the diesel fumes to the outside of the plant during the testing procedure. *ECO*

a total run of close to 1,000 units. The following month, production officially started on its replacement—the D11N.

The D11N

Starting in 1986, Caterpillar introduced the first of its new "N-series" track-type tractors in the form of the D11N. The D11N (S.N. Prefix 74Z) replaced the original D10. In 1987, the company would release another round of "N" designation tractors. The D8L became the D9N, and the D9L became the D10N. To fill the gap in the product line vacated by the D8L, a newly designed D8N was introduced. This system of track-type tractor upgrades was a bit confusing at the time. But it was necessary to make room in the product line for the model D8N. Over the years, many have assumed that the D11N was a totally new product line, and that the D10N was the upgrade to the original D10. Not true. The D10 evolved into the D11N, whereas the D10N was the improved version of the D9L—a confusing situation, to say the least.

When prototype testing was completed on the pilot D11N, the company commenced production in East Peoria in February 1986. At first glance, the D11N looked much like the D10, except for its markings. But upon closer inspection, the D11N was brimming with new features, including a new engine package with increased power ratings, and a lengthened track frame that put more track on the ground.

The original D348 engine found in the D10 was replaced with the Cat 3508 DITA, four-stroke twin-turbocharged and aftercooled V-8 diesel engine, rated at 770 flywheel horsepower (817 gross). This power plant offered a 10 percent rise in horsepower, 18 percent higher displacement, and 10 percent higher peak torque than the D10's engine. Despite the increased power, the direct fuel

injection D3508 utilized an average of 5 percent less fuel per cubic yard dozed. The turbochargers were located on top of the engine in the D11N. Twin exhaust stacks were present, as were two air cleaner intakes. The air cleaners were situated in front of the exhaust pipes on the D11N's hood. On the later D10 models, they were located behind the exhaust stacks.

The other big change on the dozer was its longer track frame. This put an additional 21 inches of track on the ground, with most added to the rear of the elevated sprocket. This helped counteract front-end rise, which increased traction and produced higher dozing and ripping force. Total track length in contact with the ground was 175 inches, compared to the D10's 154 inches. With standard track shoes, the D11N actually produced 6 percent less ground pressure than the D10, even though its overall working weight was up 5 percent.

Other significant structural upgrades of the D11N over the D10 included a 30 percent thicker roller frame tube, 22 percent thicker bottom guards, and a 22 percent increase in track pitch. The front bulkhead plate was now a one-piece structure instead of two, for greater strength. There were other improvements as well.

With more power and greater traction at its disposal, the D11N featured larger, more productive bulldozing blades. The 11SU straight blade was rated at 32.8 cubic yards, while the 11U full U-blade was capable of handling 42.2 cubic yards. In 1987, capacities on these two blades increased to 33.5 and 45 cubic yards, respectively. The 11SU blade measured 18 feet, 4 inches across, while the larger 11U had a width of 20 feet, 10 inches. Both blades were 7 feet, 7 inches in height. Both also featured more heel clearance than the D10, with sharper cutting edge angles, which enabled them to cut into tough material more easily.

After the final paint and cure, the D11R is taken to the paint detail area, where any runs, drips, or errors in the surface can be sanded out and repainted. Though not in the same league as an automobile paint application, it is one very nice finish for such a machine. The pride of workmanship on these tractors is evident at every stage in the build process in Building SS. *ECO*

D11R dozers leave the assembly building in one of two ways, by multiple tractor-trailer loads, or in this case, by reinforced railcar. When transported by rail, the D11R can go as a complete unit, minus its blade and hydraulics. Here, the ROPS bar is lowered into position and bolted down into place. Once this is accomplished, the big tractor will be set to roll. *ECO*

At the rear of the D11N could be found a single-shank ripper, a multishank ripper, and in later models, a massive impact ripper. Starting in 1987, Caterpillar Custom Products offered a giant hydraulic impact ripper. In operation, a hydraulic impactor transmitted powerful energy pulses through a specially designed forged alloy steel ripper shank, concentrating 450,000 pounds of impact at the ripper tip, 540 times a minute. The resulting shock forces fractured rock not only at the tip, but ahead of it as well. This reduced needed drawbar pull, since the tip was continuously moving through prefractured rock. The impact ripper was pulled through the ground just like the standard rippers. Only when the tractor encountered 15,000 pounds of drawbar pull, sensed in the ripper tilt cylinder, was the hydraulic hammer actuated. But the impact ripper was hard

Behind Building SS, two D11R tractors await their turn for final outfitting and shipping destinations. Most D11Rs are built as special order items, though Caterpillar likes to keep a few in inventory for quick turnaround situations, just in case a customer needs one of these giants ASAP. *ECO*

This view of the D11R highlights its suspended undercarriage design, which employs four bogie suspension roller assemblies that help the track conform more closely to the ground, providing up to 15 percent more ground contact. This gives the D11R greater traction, especially on hard and uneven ground. *ECO*

A D11R CD Carrydozer making its way through Illinois by rail is an incredible sight for both heavy-equipment and train enthusiasts. In most cases, two railcars will handle the job. If the big Cat were to travel by road, four to five semitruck loads would be required. *ECO*

In late 1996, Caterpillar introduced a second model, referred to as the Carrydozer, in its D11R product line. Designed with a special blade that actually carries a portion of the load, it is best suited for large volume dozing applications. After making its first public appearance at the September 1996 MINExpo in Las Vegas, Nevada, the D11R CD makes a repeat showing at the October 2000 MINExpo. *ECO*

on the D11N. Prolonged use of the powerful ripper would cause stress fractures in the D11N's frame, as well as its undercarriage. The impact ripper made short work of just about everything it encountered in the field, but this unfortunately included the tractor's structural components. The impact ripper option was discontinued in the early 1990s.

With all of the improvements and ripper options, the D11N weighed a bit more than the D10. In full operating trim, equipped with the U-blade, the dozer weighed in at 205,948 pounds with the single-shank ripper, 208,571 pounds with the multishank unit, and a whopping 225,950 pounds with the tail-heavy impact ripper in place.

In 1993, Caterpillar released an updated version of the D11N. This D11N (S.N. Prefix 4HK) featured an improved Cat 3508 EUI diesel, featuring Electronic Unit Injection, which provided precise fuel delivery, reduced white smoke at start-up and black smoke while running, and improved cold starting. Horsepower figures were unchanged. In the cab, the operator made use of a state-of-the-art Computerized Monitoring System (CMS), which provided feedback on machine systems and recorded

With 850 flywheel horsepower and a massive 57-cubic-yard blade, the D11R CD Carrydozer can push and carry a small mountain of earth. Weighing in at 248,600 pounds in full operating trim, it is the heaviest track-type tractor ever offered by Caterpillar to date. *ECO*

The D11R CD Carrydozer carries material inside its blade curvature for increased productivity. This increases the effective weight of the tractor, which enables it to push a larger pile of material in front of the blade. Material carried in the blade also allows operation of the dozer on steeper slopes. *ECO*

performance data for assistance in diagnosis. The operator also had a better view of the ripper from the cab, thanks to a notch just below the rear window. Weight increased again. An updated D11N in full operating trim, with a full U-blade and single-shank ripper, tipped the scales at 214,847 pounds. When equipped with the 11U-ABR abrasion blade package and a multi-shank ripper, weight jumped to 223,142 pounds.

The last version of the D11N, which featured new striping and nomenclature decals, remained in production until 1996, when the new D11R model took over the top spot in Caterpillar's track-type tractor product line.

The D11R and D11R Carrydozer

In March 1996, Caterpillar officially introduced the D11R. The new model (S.N. Prefix 8ZR) featured a host of evolutionary improvements that would continue the sales success story of the D11N, which claimed approximately 70 percent of the world's dozer sales in its size class. Though the D11R looked like the former D11N model (except for its new black operator's cab), a host of new features improved reliability and increased operator comfort and productivity. Key improvements included thicker brake plates and increased oil flow, Electric Clutch Brake (ECB) steering system with Finger Tip Control (FTC), an Improved Multiple Row Module (IMRM) radiator replacing the previous folded core unit, Caterpillar Monitoring System (CMS) with odometer, improved idler caps, upgraded equalizer bar, ecology fluid drains, and an optional air-suspension seat.

Of all the improvements, the Electric Clutch Brake (ECB) steering system with low-effort Finger Tip Control (FTC) was the biggest change. This system combined steering, machine direction, and gear selection into a control system clustered to the operator's left that could be operated with one hand, for superior operator comfort and increased productivity. Blade and ripper controls were mounted to the right of the operator's seat.

The D11R's engine, as well as its power output ratings, were the same as the D11N's.

Bulldozing blades and capacities were also unchanged from the previous model. Weight was up a bit, to 216,963 pounds when equipped with the full U-blade and single-shank ripper. Heaviest combination was the 11U abrasion blade package with multishank ripper, which tipped the scales at 225,903 pounds.

The early D11R stuck with the old D11N's power output and blade capacities. But this was merely the first installment of what Caterpillar had up its sleeve for its top-of-the-line dozer. In September 1996, at the MINExpo '96 mining show held in Las Vegas, Nevada, the company unveiled a second model of its famous track-type tractor—the D11R Carrydozer.

The D11R Carrydozer (CD) was not a replacement or upgrade of the standard D11R, but a second model type with a completely new blade design. The D11R CD (S.N. Prefix 9XR) was primarily designed for large volume production dozing in prime mine stripping and land reclamation applications. The new dozer's blade carries material inside the blade curvature for increased productivity. This feature increased the effective weight of the tractor, enabling it to push a larger pile of material in front of the blade. Material carried in the blade also allowed operation of the D11R CD on steeper slopes. Blade capacity of the Carrydozer is 57 cubic yards, 12 more than the standard D11R. The blade of the Carrydozer measures 22 feet in width, and 10 feet, 8 inches in height to the top of the rock guard.

Engineers also saw fit to install an improved Cat 3508B EUI diesel engine under the hood. The B-model of this engine meets tougher governmental standards concerning lower sound and emissions. The engine block has added internal ribbing around a new larger camshaft bore. Improved coolant passages in the cylinder heads, plus electronic unit injectors with reduced internal restrictions, also improve performance measurably. A new two-pass Advanced Modular Cooling System (AMOCS) kept engine temperatures under control. The new D11R CD now boasted a power output of 850 flywheel horsepower (915 gross) at 1,800 rpm. In

The D11R CD's power output of 850 gross horsepower makes it an extremely productive machine in mining operations the world over. At full throttle, the D11R CD is awesome in sight and sound, as the noise of its 34.5-liter Cat turbocharged and aftercooled V-8 echos across the landscape. *ECO*

With its 850 flywheel horsepower and 35.5- and 45-cubic-yard bulldozing blades, the D11R was designed to take on the biggest dozing jobs within today's large open-pit mining operations. With an operating weight of 230,100 pounds, nothing gets in the way of this big Cat. This D11R is shown at work in northern Alberta, equipped with the 35.5-cubic-yard SU (semiuniversal) blade and single-shank ripper. *ECO*

operation, the D11R will run out of traction long before it runs out of power. In fact, with an extremely full blade, the tracks will slip and crawl as they constantly bite into the ground, but the engine never misses a beat, never strains, and is never wanting for more power. The engine sounds the same at the beginning of a dozing run as it does when it reaches the end.

Other performance-enhancing improvements to be found on the D11R CD included a beefed-up frame, extra-heavy-duty single-shank ripper carriage, and new low-effort electronic dozer control, which allows the operator to control all of the dozer functions with one hand. The ripper and implement controls were now all electronic as well. An optional Computer-Aided Earthmoving System (CAES) guided the operator's dozing patterns with the aid of Global Positioning System (GPS).

As one might expect, all of these new features added a bit of weight to the tractor. Weighing in at 239,550 pounds in full operating trim, it is even heavier than the old D11N equipped with the giant impact ripper. At the time of its introduction, the Carrydozer was only offered with a single-shank ripper. If the ripper is not required, then a 29,620-pound counterweight is installed instead to maintain the tractor's overall balance.

The D11R CD had limited availability throughout 1997, as it went through its test phase. Full production officially got under way by April 1998.

In August 1997, Caterpillar released an updated version of the standard D11R, which featured many of the improvements first seen on the new Carrydozer model a few months earlier. The standard D11R (S.N. Prefix 9TR) now featured the 850-flywheel-horsepower engine, including

the new cooling system, stronger main frame, and improved undercarriage and dozer structures. Standard operating weight of the dozer with full U-blade and single-shank ripper was now 222,133 pounds, with the 11U-ABR blade and multishank model coming in at 228,598 pounds.

Entering the twenty-first century, the D11R and the D11R CD are the most powerful and productive track-type crawlers ever offered by Caterpillar. In late 1999, Caterpillar added further upgrades and features to the big dozers to make them even more productive. The D11R (S.N. Prefix 7PZ) and D11R CD (S.N. Prefix AAF) now featured an optional environmental cab with a host of interior changes, including a new comfort seat. The latest in controls, including the Vital Information Management System (VIMS) and Computer-Aided Earthmoving System (CAES) can also be found on the dozers. With CAES, the operator is given an in-cab, real-time work plan, which is continually updated to indicate where to cut and fill. Using on-board computers, sophisticated software, and GPS, it virtually eliminates the need for maps and grade stakes. Also, the Carrydozer now could be optioned with a multishank ripper. These changes again added weight to these high-end production dozers, with the standard D11R now coming in at 230,100 pounds, and the D11R CD at 248,600 pounds, the heaviest of any Caterpillar track-type tractor. In addition to the other enhancements mentioned, the Carrydozer would receive further improvements in October 2000, with the addition of a strengthened main frame and new tag link design, which now attached the link to the blade instead of the push arms. This made sure that blade side loads were applied directly to the main frame to ensure a long connector life.

The D11R is usually equipped with one of two ripper options, a single-shank or multishank configuration. This unit is equipped with the single shank, which is designed for severe ripping applications, especially in hard-packed and rocky working conditions. If the ripper is not needed by the customer, an 11,000-pound counterweight is required in its place to maintain the tractor's overall balance. *ECO*

CATERPILLAR

The multishank ripper option on the D11R is best used for general ripping operations. It can be set up with one, two, or all three shanks in place. Its parallelogram linkage, with hydraulically variable pitch, can angle the tips of the shanks for the best ground penetration. This D11R, at work in the Powder River Basin coal mining area of Wyoming, is equipped with the multishank ripper and 45-cubic-yard dozing blade. It also features optional black antiglare paint scheme and single upper track carrier rollers. *ECO*

The D11R makes life for the operator easier as well, with its use of Caterpillar's Finger Tip Controls (FTC). FTC combines steering, machine direction, and gear selection all in a single hand control mounted just to the left of the operator's seat. This control module can be adjusted up and down and fore and aft for maximum comfort for the operator. *ECO*

The Competition

The Caterpillar D11R currently holds 80 percent of the world track-type tractor sales in its size class. But that doesn't mean that there are not a few other large dozers with the muscle to take on the big Cat. Actually, there are four contenders, all designed and built by Komatsu. They are the D475A-3, the D475A-3 SD Super Dozer, the D575A-3 SD Super Dozer, and the D575A-3 SR Super Ripper. The D475A-3 series is closest in power and mass to the D11R. The D575A-3 SD/SR is quite a bit larger than the dozer built in East Peoria, Illinois. In fact, the Komatsu D575A-3 SD is the largest and most powerful production track-type tractor in the world.

The Komatsu D475A-1 series was introduced in 1987 as the principal rival to the D11N. Today's D475A-3, introduced in late 1998, comes in two model types: The standard model is rated at 860 flywheel horsepower, 10 more than the D11R. Its largest bulldozer blade is a 45-cubic-yard full U-type, the same as the D11R's. Overall working weight equipped with the U-blade and single-shank ripper is 228,660 pounds. The D475A-3 SD is Komatsu's answer to the D11R Carrydozer. This dozer has the same power ratings as the standard model, but comes equipped with a larger 58.9-cubic-yard high-volume dozing blade, which is just 1.9 cubic yards more than the D11R CD. No ripper option is offered for the D475A-3 SD. Working weight is approximately 231,480 pounds, some 17,120 pounds less than the current Carrydozer. In many vehicles, extra weight is a bad thing. But in dozers just the opposite is true. Extra weight translates to better traction in most cases, just as long as there is enough horsepower and torque available. The D475A-3 series is a formidable dozer. It meets or exceeds the D11R tractors, point for point. The big Komatsus also offer many of the state-of-the-art technological features found on the D11R. Though its undercarriage is very advanced, it still relies on a more conventional low sprocket drive design, as compared to Caterpillar's elevated sprocket system.

At the top of the dozer food chain are the Komatsu D575A-3 SD and SR track-type tractors. These giant dozers first started field-testing in North America in 1989, with production officially getting under way in 1991. In 1995, the first D575A-2 SD Super Dozer was introduced. In 2001, it was replaced by an upgraded D575A-3 SD. At first glance, the D11R dozers simply seem outgunned by these monster Komatsu machines. And on paper, the numbers are hard to argue with. The D575A-3 SD pumps out 1,150 flywheel horsepower, wields a massive 90-cubic-yard blade, and

This D11R track-type tractor is one of two such 850-flywheel-horsepower units operated by McAninch Corp. of West Des Moines, Iowa, that are used for pull-type scraper work. The D11R was originally ordered from the factory minus all of its hydraulic blade controls, and is shown here attached to a modified 651E scraper unit with a custom built dolly. *Caterpillar, Inc.*

breaks the scales with its 336,483-pound overall working weight. Incredible statistics. On the downside, is no ripper option and a substantially higher price tag, in some cases, almost 2 1/2 times as much as a standard D11R. The second model is the D575A-3 SR Super Ripper. Its power output is listed at 1,050 flywheel horsepower, it comes equipped with a 70.8-cubic-yard full U-blade, and it weighs in at 289,030 pounds. As the name of the machine implies, the D575A-3 SR comes equipped with a massive single-shank ripper capable of a maximum digging depth of 6 feet, 9 inches. While quite impressive, these giant Komatsu dozers are very specialized machines and are designed for jobs especially suited to their size and specialty. The D11R dozers are more versatile than these giants, but then that is why Komatsu offers the D475A-3.

Komatsu has always been the thorn in Caterpillar's side concerning its D10/D11N/D11R big dozer offerings. But this competition keeps Caterpillar engineers pressed to keep the company's machines the most effective tools they can be. Never one to rest on its laurels, Caterpillar keeps pushing the technological boundaries of what can be done and implemented in an ultralarge dozer. That is why the D11R is the best-selling mining dozer in its size class in the world.

The 797 was designed to be loaded by today's largest cable and hydraulic mining shovels. Many of the cable shovel designs offer bucket capacities in the 100-ton range, enabling them to load the 797 in four very quick passes. *ECO*

CHAPTER TWO

The World's Largest Mechanical-Drive Truck

797 HAUL TRUCK

Caterpillar is considered one of the leading manufacturers of off-highway rear-dump trucks in the world today. But this hasn't always been the case. In fact, the rear-dump hauler was only added to the company's product lines in late 1962. There are numerous reasons why Caterpillar took so long to get into the haul truck business. During the 1940s and 1950s, the heavy-duty off-highway truck market was dominated by companies such as Euclid, Dart, and Mack. These manufacturers had been building trucks for decades and knew the ins and outs of the marketplace—the same way that Caterpillar understood the track-type tractor market. But Caterpillar dealers were requesting a broader product line to offer their customers one-stop purchasing of their construction equipment needs. The company did offer Athey-built rear- and bottom-dump trailers for various one- and two-axle Caterpillar tractors originally designed for scraper use. But they were far from ideal and simply not in the same league as a true off-highway truck.

Pictured in November 1962, this is the preproduction 769X4, the design that would get the go-ahead to start full assembly-line production in January 1963 as the 769, Caterpillar's first true rear-dump off-highway hauler. *ECO Collection*

Management knew that the dealers were right. And as the sales of quarry and mining trucks continued to increase in the industry, Caterpillar decided it had no choice but to enter the fray or be left behind for good. The potential for huge profits from off-highway trucks also wasn't lost on some of the company's closest rivals. LeTourneau-Westinghouse had recently introduced its first "Haulpak" LW-30 mining truck in 1957. This was followed by the LW-80 Coal Hauler later that same year. In late 1958, Caterpillar management authorized development of the company's first off-highway hauler.

Caterpillar had just introduced its first rubber-tired front-end loader, the model 944A Traxcavator, in 1959. This was the same year that the first experimental prototype rear-dump hauler would be completed. In July, engineers produced a 28-ton-capacity truck featuring larger rear tires than the fronts, mounted on single wheel rims, one per side. It also was equipped with an air-oil strut suspension system. This was the first step in what would become Caterpillar's first off-highway truck, the model 769.

Not long after the first 769X1 prototype hauler was built, a second pilot truck, 769X2, was constructed in 1959. This unit was built without a rear air-oil suspension system with the hope that the rear tires themselves would provide sufficient spring-rate for an adequate ride and as a cost-reducing factor. But it didn't work. Lessons learned from these first crude efforts were put to the test in the third prototype truck, 769X3, in January 1962. This effort looked far more sorted out. Tire sizes were the same all the way around, with dual tire wheel mounts in the rear. The front-end body sections and cab were fabricated out of hand-laid fiberglass. Capacity of a newly designed dump box was rated at 35 tons. By November 1962, the fourth prototype was finished and ready for testing. Referred to as 769X4, it would be the design of choice to enter production. Looking much like the 769X3 unit, the 769X4 sported a different nose and radiator grille

Caterpillar entered the 240-ton capacity class of mining trucks in 1991, when it introduced its model 793. With its mechanical-drivetrain layout and 240-ton payload, the 793 series would soon become the world's best-selling hauler in its tonnage class. Pictured is the first 793 unit built, undergoing tests at a mine site adjacent to the Cat Arizona Proving Ground in 1991. *Caterpillar, Inc.*

In 1992, the 793 was upgraded into the 793B series. Though the models looked virtually the same, a host of improvements increased reliability and productivity of the big mining hauler. Most notable was the improved Cat 3516 EUI (Electronic Unit Injection) diesel engine, which reduced smoke at start-up and increased fuel economy. *ECO*

design. This rounded corner look, with round hole steel mesh radiator guard front, would establish the look of Caterpillar's mechanical-drive hauler models for the rest of the decade and into the early 1970s. Full-scale production of the model 769 finally got under way at Caterpillar's Decatur, Illinois, assembly plant in January 1963.

In 1961, Caterpillar engineers started to set their sights on a much larger haul truck intended for use in hard-rock mining applications, since the smaller 769 was basically designed for use in large construction and quarry operations. To help put the project on the right track, the company hired Ralph H. Kress in 1962 as manager of truck development for the Research and Engineering Department of Caterpillar Tractor. Kress was the former manager of truck development for LeTourneau-Westinghouse and designer of the first trendsetting Haulpak truck. After numerous market studies, the company decided to make the hauler a diesel-electric-drive truck with a targeted 75-ton capacity. This hauler would be the model 779. The 779 was designed and built alongside two other experimental prototype hauler concepts, identified as the models 783 and 786, which also utilized the electric-drive system.

The prototype Caterpillar 783 was rated as a 100-ton-capacity hauler. Its unique features included a three-axle design, with only the middle axle driven. The truck was also fitted with a side-dumping body, capable of unloading from the right or left side. The 783 started testing in August 1965. Only one chassis was built, utilizing two different dump body designs.

Largest of all of the Caterpillar diesel-electric-drive creations was the giant 786 Coal Hauler. Introduced in October 1965, it shared many of the same drivetrain components found in the 779 and 783 haulers. The 786 utilized two drive systems, one in each tractor, powering the front axle on each unit. With two Cat D348 Diesels, the giant bottom-dump was capable of generating

At the 1996 MINExpo, Caterpillar introduced its 793C off-highway hauler. Again, looking much like its predecessors, the 793C was equipped with the new clean-burning Cat 3516B EUI diesel engine. Power output of this unit is listed at 2,166 flywheel horsepower (2,300 gross). Overall loaded gross weight of the truck is 846,000 pounds, with a payload capacity of 240-plus tons. *ECO*

Caterpillar offers a multitude of Mine Specific Design (MSD) bodies that can be custom matched to a customer's specific mine-hauling needs. From overburden to coal, Cat designs the right MSD to get the most productivity possible out of its haulers. This 793C is equipped with a lightweight MSD, which raises the nominal payload capacity of the truck to 270 tons. This one has a bit more on it for demonstration purposes. *ECO*

1,920 flywheel horsepower. The giant hopper was rated at 240 tons, making it the world's largest hauler in its day. Five units were eventually built—the original prototype and four limited production sets—in 1968.

By mid-1965, the first rear-dump pilot 779 was ready to start prototype testing. By October 1966, the 779 was given the go-ahead to start full production at the Decatur assembly plant. But early problems with the 779 kept delaying production. The 779 frame was displaying numerous stress fractures in the field. After engineers redesigned the frame, the company officially released the 779 in July 1967. At this point, the truck was still rated as a 75-ton-capacity hauler, but in 1968, after further upgrades, including a redesigned dump box, the payload rating was increased to 85 tons.

But the problems with the hauler kept on coming. The frame, the electric drive, and the suspension system were key sore points. Finally, Caterpillar officials had had enough, and to cut any further loses, they canceled the 779 along with all of the other electric-drive trucks. The last 779 came off the assembly line in May 1969. To eliminate any warranty or liability problems with the trucks, all of the 779 haulers in the field were reacquired by Caterpillar starting in late 1969 and throughout 1970. All were then dismantled. Despite the program's failures, Cat engineering gained tremendous knowledge concerning the building of mining trucks.

With the end of the diesel-electric drive programs, Caterpillar set its sights once again on expanding the mechanical-drive hauler line. Proven industry-accepted tonnage size classes for the trucks would be priority one, translating to larger sales volume for the company. In 1970, the company introduced the 50-ton-capacity model 773. It looked much like its little brother, the 769B, which had been upgraded in late 1966. The 773 packed 600 flywheel horsepower under the hood. Off-highway haulers in the 50-ton class, like the 35-ton category, were very popular in the earthmoving industry. Large construction and earthmoving contracts, as well as quarries and

smaller mining operations, utilized haulers in this size class in tremendous numbers. For the time being, Caterpillar was going to let the larger mining hauler market be serviced by manufacturers such as KW-Dart, Euclid, Unit Rig, and WABCO. But this deferment was to be only temporary. As the 769 and 773 hauler product lines began to dominate their respective size classes in the industry, Caterpillar set its sights on the next most popular tonnage class, the 85-ton-capacity off-highway hauler.

In 1974, Caterpillar introduced what would be one of its most popular off-highway haulers, the 777. The Cat 777 filled the void in the hauler line left vacant after the termination of the diesel-electric-drive 779 program. The 777 was powered by a 870-flywheel-horsepower diesel engine, mated to a seven-speed fully automatic transmission. Other niceties included oil-cooled brakes, an all-new ROPS cab, and a totally new look for the front radiator housing and decking. This design was so well thought out, that it remains the basic look for Caterpillar haulers even today. With its 85-ton capacity and modern design, the 777 quickly achieved what it set out to do, to give the customer a superior hauler with unmatched productivity for its size class. In so doing, it became the dominant force in the 85- to 100-ton-hauler capacity class.

With three successful off-highway trucks under its belt, the company set its sights on far bigger game. The larger mining truck industry had for years been dominated by diesel-electric-drive haulers in the 150- to 200-ton size classes. As more mining operations around the world upgraded their haulage fleets with trucks in these size classes, it now made financial sense for Caterpillar to also field off-highway hauler designs in this market. But with a difference. The Caterpillar offerings would utilize mechanical-drivetrain designs, in contrast to the diesel-electric-drive creations built by the other manufacturers.

The first of the company's larger mining haulers was the 785. Officially released to the mining industry in 1984, the 785's design was in keeping with the 777 series, but larger. The 785

Caterpillar's largest mining truck to date is its incredible 797. Introduced in late 1998, the giant hauler is the world's largest mechanical-drive mining truck. *ECO*

The heart of the 797 is its Caterpillar 3524B EUI diesel engine. This engine was developed by combining two Cat 3512 engines in-line by means of an innovative flexible coupling system, which allows both units to utilize a single common crankshaft. Power output of this massive 24-cylinder, 7,143-cubic-inch (117.1-liter) engine is 3,211 flywheel horsepower (3,400 gross) at 1,750 rpm. The torque rating is a mind-numbing 12,170, foot-pound. Overall engine module weight tips the scales at approximately 40,000 pounds. *ECO*

had a nominal payload rating of 130 tons, yet a maximum load capacity of 150 tons was possible with the larger tire option. Up front was an extremely reliable V-12 Cat 3512 diesel engine, rated at 1,290 flywheel horsepower (1,380 gross). A six-speed, electronically controlled automatic transmission completed the package. From the start, the 785 was a winner out in the field. And like other popular Cat off-highway haulers, it too would soon dominate its capacity class, becoming the number one choice in the 130- to 150-ton truck market.

Not long after the introduction of the 785 came an even larger model, the 789. Released in 1986, the 789 was everything the 785 was, but in a slightly larger package. The 789 carried a standard payload rating of 170 tons, with a maximum 195-ton capacity when equipped with the larger tire and wheel option. Power was supplied by a 1,705-flywheel-horsepower (1,800 gross) Cat 3516, V-16 diesel engine, mated to a six-speed electronically controlled, fully automatic powershift transmission mounted to the rear drive differential, just like the 785. Though the 789 is larger than the 785, it is hard to tell one from the other at a distance. The quick way is to count the front ladder steps. The 785 has five steps, while the taller 789 has six. The Cat truck soon equaled and then surpassed its diesel-electric-driven competition in the 170- to 200-ton size class, by delivering 20 to 25 percent more material

Here a 797 chassis, complete with its engine package in place, slowly makes it way down the assembly line. As one might suspect, building anything as large as the 797 is a hands-on endeavor, unlike the automotive industry, which automates the process with more robots. *ECO*

Painting a truck the size of the 797 is a daunting task. Here, three workers at the Decatur, Illinois, Cat assembly plant apply the final coat of paint to an outfitted frame of the 797. Approximate paint time will be 1 hour and 45 minutes. Overspray from the paint process is pulled out of the air through venting in the floor. It will take 15 gallons of primer and 15 gallons of finish paint to complete this portion of the truck. Overall, a completed 797 will require just over 110 gallons of paint. *ECO*

The massive radiator of the 797 is built as a separate component module. The radiator fan is hydraulically driven and electrically controlled, which decreases fuel consumption and noise. It has a capacity of 315 gallons of coolant. *ECO*

Like the radiator, the rear axle differential and wheel stations' subassembly is temporarily installed, along with the transmission, so the complete drivetrain can be tested within the plant. Here, a portable filtering system is circulating differential fluid under pressure throughout the rear subassembly. This filters any possible stray metal particles or contaminants out of the system that might still remain, ensuring higher reliability in the field. *ECO*

for every gallon of fuel burned when compared to these other trucks.

It was becoming clear to the industry that Caterpillar was methodically targeting the most popular off-highway size classes one at a time, instead of all at once like so many builders before it. Though this added to the time it took for the company to build up a substantial off-highway product line, the trucks were so well engineered, they became sales leaders practically overnight. Longtime truck builders such as Dart, Mack, and Euclid were simply overwhelmed by the sales success of these Caterpillar machines. Another advantage that Cat had over its competition was the use of its own diesel engine designs. All of the other manufacturers relied on Detroit Diesel and Cummins for their diesel power.

At the time the 789 series was released, the mining industry was on the verge of stepping up

to the 240-ton-capacity size class. Several companies had anticipated the move up to substantially larger haulers during the 1970s, with trucks like the 250-ton-capacity Peerless VCON 3006 and the 200- to 235-ton WABCO 3200, both in 1971, and the monstrous 350-ton Terex 33-19 Titan in 1974. But those giant machines were ahead of their time in terms of industry needs.

In the 1970s, it took slow revving, high-horsepower locomotive diesel engines to power these trucks. Those engines came with a weight penalty. Higher revving, lighter weight diesel engine designs of the time were simply not powerful enough to move such massive vehicles. This situation started to change in the 1980s when WISEDA (now Liebherr) broke the 220-ton-capacity barrier with a two-axle production hauler referred to as the KL-2450. Introduced in 1982, it featured a high-revving 1,800-flywheel-horsepower Detroit Diesel 16V149TI engine. By 1985, WISEDA raised the capacity bar even further with the release of the industry's first 240-ton-payload truck. This payload capacity was looked at with the most favor in the mining sector. New loading shovels produced by P&H, Marion, and Bucyrus-Erie, with bucket capacities of 80 tons, could load such a hauler in three passes. This would be the benchmark followed by the mining industry for the next few years.

Soon after the release of the 240-ton WISEDA, other manufacturers stepped up to the plate with their truck offerings, most notably the Dresser (formerly WABCO) 830E Haulpak and the Unit Rig Lectra Haul MT-4000, both in 1988. Rated at 240-tons payload, the Unit Rig and Dresser diesel-electric-drive haulers soon established solid sales records in large mining operations the world over, especially the 830E Haulpak, which was the number one choice in this size class. But Caterpillar engineers by this time were well under way on their own 240-ton-capacity off-highway hauler.

In 1987, Caterpillar formed the Mining Vehicle Center (MVC) at its Decatur, Illinois, assembly

The giant radiator module is temporarily installed so all systems can be brought up to power to fully test the drivetrain on the 797 before shipment. Once everything checks out, the radiator is removed and shipped separately, to reduce the weight of the chassis. *ECO*

CATERPILLAR
797
797
201

plant. Officially announced in 1988, its main purpose was to develop and manufacture a new generation of mining machines significantly larger than those currently built by the company. These would include a new wheel loader (994), a motor grader (24H), a larger hydraulic excavator (5130), and a 240-ton-capacity truck to be known as the 793 series. The 793 hauler, as well as the rest of the new machines, would utilize Caterpillar mechanical drivetrains. All of these designs were produced using computer-based analytical tools to reduce the time needed to bring them to market. By utilizing a process referred to as "simultaneous engineering," development times were reduced measurably. The new MVC started with just three employees in 1987, but quickly expanded as new concepts were approved for manufacture. In 1989, renovation work started at the Decatur plant on the production facilities that would handle the building of these larger concepts. By late 1990, construction of the first pilot 793 hauler was under way in Decatur.

Everything about the 797 is huge. Rated at a nominal 360-plus tons, the 797 is one of the largest off-highway haulers in this size class in full production today. Here it is shown next to a Ford F-150 pickup truck, which gives you an idea just how immense the 797 really is. *ECO*

The 793 was designed entirely around a mechanical drivetrain concept, in contrast to the competition's purely diesel-electric-drive machines. Though many in the industry thought that a powershift transmission could not be built for life in a 240-ton hauler, Caterpillar engineers proved them all wrong with the 793. When production officially started in January 1991, the 793 instantly became the world's largest mechanical-drive truck available to the mining industry.

The 793 was a new design, and not just a beefed-up 789 series. Though both trucks had the same general appearance, the 793 was far larger, and incorporated an operator's staircase in front of the main radiator housing, which made getting on and off such a large hauler much safer. The 793 was powered by a Cat 3516 diesel, the same type found in the 789 series. But in this application, output was increased to 2,057 flywheel horsepower (2,160 gross). Transmission choice was an ultrasophisticated, mechanically driven, electronically controlled, six-speed powershift unit. This was the first time a transmission of this size had ever been installed in this class of mining truck. It was the heart of the 793, as a problem here would shut down the whole vehicle. Aside from a few minor bugs, the unit worked as designed. All shortcomings were eventually exorcised from the transmission by the time it became the 793B series in 1992.

The 793 series soon established itself as the mining industry's number one choice in the

For shipping, the rear subassembly, which consists of the banjo housing, differential, and rear wheel stations, is loaded onto a heavy-duty lowboy trailer. Weighing in at approximately 82,300 pounds, the rear axle subassembly takes up all the room on the trailer. Six or seven semitrailer trucks or five railcars are required to transport the main chassis components of a finished 797 out of the Decatur facilities. Tires ship from their point of manufacture and require two semitrailer trucks. The dump body ships in four loads from its point of fabrication. *ECO*

240-ton-capacity haul truck class. By the mid-1990s, it surpassed the former diesel-electric champion, the Komatsu-Dresser 830E. In September 1996, the 793B became the 793C. Though initially rated at 240 tons, by the year 2000, it received a rating of 240-plus tons, depending on tire and dump body configurations.

Though Caterpillar was clearly besting the 830E in the marketplace, Komatsu, which now controlled 100 percent of the Komatsu-Dresser joint venture company, decided to up the ante with a new truck identified as the 930E. Introduced in 1995 with a 310-ton capacity, it would be the first of the so-called ultrahauler class of two-axle mining haulers to go into full production. By 1997, this rating had increased to 320 tons. The 930E also boasted a very efficient AC diesel-electric drive system, codesigned with General Electric. Before this, all production electric-drives in the off-highway marketplace were DC in nature. Large mining operations in North and South America looking for larger haulers to increase their profitability found the 930E package hard to resist. Mines that utilized 80- to 85-ton-capacity shovels to load their 240-ton hauler fleets in three passes, found that it took only four passes to fill a 930E with 320-plus tons of material. More tons per trip meant more money in the mine owners' pockets.

Caterpillar viewed the ultrahauler truck class as the next logical step in large-scale mining operations worldwide. But this transition would not happen overnight. To make the ultrahauler economically viable, the infrastructure of the mining operation would have to match the truck's capabilities, as well as its maintenance requirements. Larger shops and wider haul roads would be needed. Plus, the main loading shovels, whether cable or hydraulic in nature, had to match up with the hauler's capacities so as to not underfill or overfill them and to load in as few passes as possible. This has been a common dilemma in the mining industry for decades—which comes first, large loading shovels or big haulers? Too many times in the past, large cable shovels were introduced by manufacturers in the hopes that the truck builders would follow with designs of trucks that were a suitable match. And likewise, truck manufacturers have built experimental haulers seeing if the shovel engineers would follow suit. Sometimes they did, sometimes they didn't.

Throughout the late 1990s, Caterpillar managers, sales personnel, and engineers studied the haul truck market. They looked at all viable drive system configurations, and gathered an enormous amount of input from their customer base as to what they would like to see in a large capacity

hauler. After all, Caterpillar was not interested in building a truck that nobody wanted. After months of intense research, Caterpillar management decided that a truck larger than the 793 series would be needed in the future to meet its customers' needs. It would have to be in a payload class of 340 to 360 tons, and utilize a mechanical-drivetrain configuration. With design perimeters set, work could now commence on the monstrous Caterpillar off-highway truck to be known as the 797.

Work on the 797 project began in earnest in early 1997 at Caterpillar's Mining & Construction Equipment Division in Decatur, formerly known as the MVC. The truck was designed with 3D computer technology, utilizing advanced finite element and solid modeling software. By using these various computer-aided design (CAD) programs, entire components and structures could be tested for fit and compatibility with each other before a single piece of iron was ever machined, saving both time and money.

One of the first issues that needed to be addressed was the engine type that would power the mammoth hauler. In the past, the company had a variety of diesel engines on hand to meet the needs of any of its equipment designs. But this situation was different. The 797 was going to require a power plant that could produce well over 3,000 horsepower. The largest engine in the 3500 series was the 3516B, V-16 diesel found in the 793C. But even this giant hunk of iron was too small for the likes of the 797. Though Caterpillar's 3600 series of industrial and marine engines could definitely provide the necessary power, their slow revving nature and heavy weight made them better suited to the locomotive and shipbuilding industries. After much discussion, Caterpillar engineers came up with a novel idea for the truck's engine. Instead of one large engine, Cat would combine two of its 3512B diesels in line, connected by an innovative flexible coupling system. This design used a single common crankshaft. In this fashion, the engines would no longer function as two separate units but as one. This new diesel would be known as the 3524B.

It is early in the morning on August 28, 2000, and the main chassis for the 797 hauler that will be on display at the MINExpo show is loaded onto a heavy-duty 13-axle (9 for the trailer and 4 for the tractor) beam deck trailer. This is the single largest module load of the truck that can be transported by road because of weight and height restrictions. *ECO*

As the early morning fog begins to lift, the long 13-axle heavy-haulage tractor-trailer rig slowly slips out of the Decatur plant with the 2000 MINExpo 797 truck chassis, beginning its trip to Las Vegas, Nevada. *ECO*

The 3524B series is a 24-cylinder, 7,143-cubic-inch (117.1-liter), four-stroke, turbocharged and aftercooled diesel engine, featuring four single-stage Garrett 60-Series turbochargers, Electronic Unit Injection (EUI), and an ADEM II electronic controller. It is rated at 3,400 gross horsepower and 3,211 flywheel horsepower, both at 1,750 rpm. The torque rating is a ground-pounding 12,170 foot-pound. The engine is connected via driveshaft to a rear-mounted, seven-speed, electronically controlled, automatic powershift transmission, attached to the front of the main differential housing. Single-lever shift control provides shifting in all gears, with each shift individually modulated for maximum smoothness. Top speed of the truck in seventh gear is 40 miles per hour on level ground with a full-rated load on board.

The modular differential is rear mounted, which greatly improves access for maintenance. It is pressure lubricated for greater efficiency and longer operating life. A hydraulically driven lube and cooling system operates independently of the speed of the truck and pumps a continuous supply of filtered oil to each final drive in the differential. This rear-axle assembly is the largest ever designed for a production piece of Caterpillar equipment.

The load-bearing frame design of the 797 differs from that of the 793C hauler in that it is made up entirely of mild steel castings. Nine major castings are machined for a precise fit before being joined by robotic welders. The company's smaller 793C uses a mixture of box section steel structures, with steel castings in critical areas. The 797 frame design uses far fewer welds and creates a structure that is incredibly strong, suitable for supporting the kinds of loads the truck will face on a daily basis. It takes the robotic welding system approximately 18 hours to weld a single frame. Of all the welds on the 797's frame, 96 percent are made by robot and 4 percent are manually done. When finished, the frame alone will weigh in at just over 91,000 pounds.

Another key factor in the design of the 797 is its use of 63-inch wheel rims. Previously, the largest rims available were 57 inches, the industry standard for haulers over 200 tons in capacity. Caterpillar worked closely with Michelin on the development of the tires that would support the massive truck. The tires, new 55/80R63 radials, were state of the art for haul trucks. Each tire was 12 feet, 9 inches in diameter, with a 55-inch tread width, and weighed in at 9,880 pounds.

Stopping a hauler the size of the 797 requires huge oil-cooled, multiple disc brakes front and rear. The brake discs are 42 inches in diameter, with 10 each on the front wheels and 15 per corner at the rear. On grades, the brakes are controlled by the Automatic Retarder Control (ARC), which maintains an optimum engine rpm and oil cooling, resulting in faster downhill speeds. The Traction Control System (TCS) uses the rear brakes to

Once assembled, the 797 was put on display within the massive convention center of the Las Vegas Hilton for the MINExpo 2000 show dates, October 9 through the 12. Here, show attendees gaze up in astonishment at the bulk of the monstrous mining truck. The 797 on display was fitted with new, wider Michelin 58/80R63 tires mounted on new 44-inch rims. It was also equipped with a special lightweight MSD body. *ECO*

improve traction by regulating the flow of cooling oil for constant retarding capability and superior truck performance on downhill grades.

The suspension system utilized on the 797 is not that much different from those found on the company's other mining haulers, just a good bit larger. The system is independent at all four corners, and utilizes self-contained oil-over-nitrogen struts, providing the operator with a smooth, well-controlled ride.

On September 29, 1998, Caterpillar unveiled a truly astounding piece of engineering work—the 797—at its Decatur, Illinois, assembly plant. The 797 is simply a huge hauler, designed to carry a payload of 360-plus tons of material. From concept to finished prototype, it took Caterpillar only 18 months to make the 797 a reality. During this time, over 300,000 design hours and 50,000 lab test hours were expended by the hauler's design teams in total. The 797 utilizes seven on-board computer systems with "flashable" chips to control and monitor all of its vital functions. Caterpillar's other mining haulers shared many design and component aspects with each other to varying

797
109

This 797, operating in northern Alberta in October 2001, is equipped with a special MSD lightweight body. The weight savings in the body translates directly to increased payloads. In this case, the nominal payload rating of this 797 is 380-plus tons. By the end of December 2001, approximately 78 of the giant haulers had been put into service worldwide. *ECO*

Another example of a special MSD body is this gargantuan 588-cubic-yard coal design. Designed by Caterpillar engineers and fabricated by WOTCO of Casper, Wyoming, they were built for hauling coal by 797s in the Powder River Basin of Wyoming. Average load carried by these trucks is over 380 tons. The first of these coal MSD bodies went into service in 2000. *ECO*

degrees. But the 797 is all new, and shares none of its major systems with its little brothers in the product line.

Everything about the 797 is massive. The truck measures 47 feet, 8 inches in length, and 21 feet, 5 inches to the top of the cab. But it is the truck's overall width that is most impressive, measuring 30 feet to the outside of the rear tires. Simply enormous. (For comparison, the Cat 793C has a width of *only* 24 feet, 4 inches.) Total operating weight of a fully loaded 797 is a whopping 1,345,000 pounds. To put the size of the truck in perspective, a fully loaded 797 is nearly 40 percent heavier than a fully loaded Boeing 747 jet airliner. The size of the hauler equates to a 4,500-square-foot, three-story home. Even the fluid capacities are vast. Standard fuel capacity is 1,000 gallons, with optional 1,600- and 1,800-gallon tanks available. Typical fuel consumption per working day averages 1,500-plus gallons of diesel. And you thought your sport-utility vehicle used a large amount of fuel! You could drive a standard pickup truck 30,000 miles on the amount of fuel burned in the 797 in a single day. The cooling system requires 315 gallons of coolant. Other fluids and oils utilized by the hauler, such as those found in the hydraulic systems, differential, brakes, and crankcase, add up to another 1,514 gallons.

After the unveiling of the first 797 chassis at the Decatur plant, the hauler was broken down and shipped to the company's Arizona Proving Ground, just south of Tucson. In November 1998, the first 797 was joined by a second prototype truck at the proving grounds. Transporting a vehicle the size of the 797 involves some serious logistics, along with a large stack of shipping permits and paper work. The main chassis, engine module, and rear differential all ship from the Decatur plant either by six or seven semitrailer trucks, or by five railcars. The tires are shipped from their point of manufacture and require two semitruck loads. The standard dump body is also shipped from its main manufacturing source and needs four semitrailer trucks. In total, 12 to 13 semi loads are needed to supply all of the components to build a single 797.

The production development phase of the 797 program started with approximately 20 units being built and put into service in various large mining operations, first in North America, then in South America. Trucks three and four, which were the first of the "production" trucks, were sent to the giant Bingham Canyon copper mine in Utah

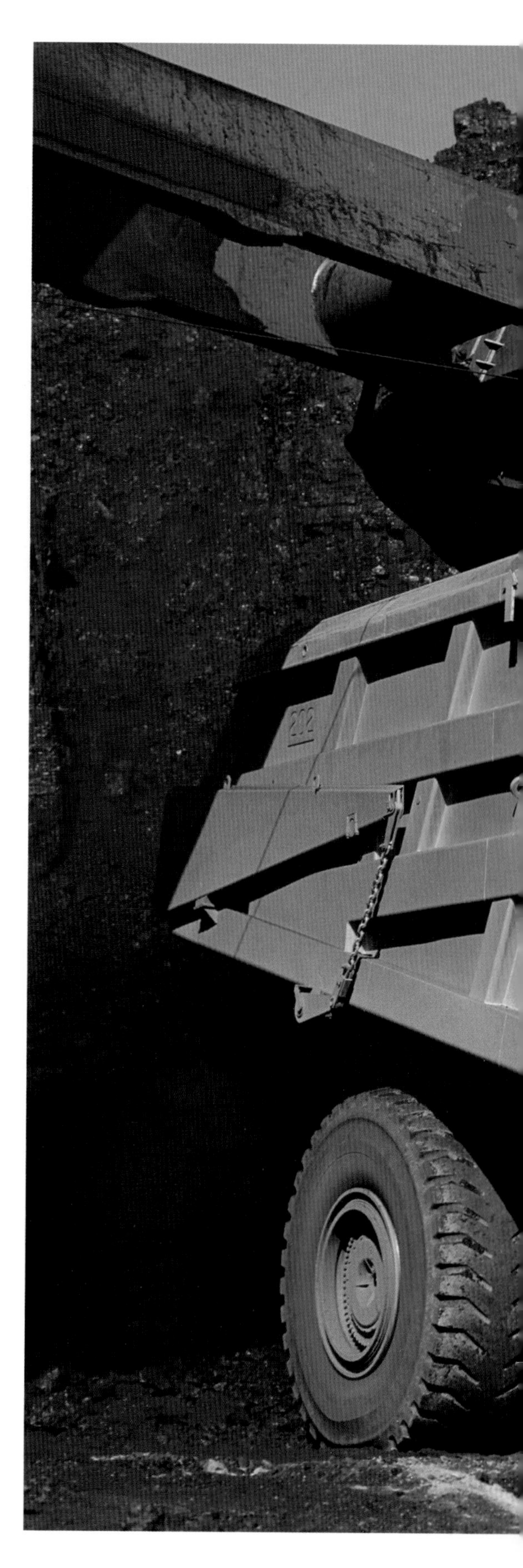

CATERPILLAR
CAT
797
202

A 797 equipped with the MSD coal body slowly makes its way out of the main pit with a full load on board. Because coal is a much lighter material, it takes a far larger volume of it to reach the 797's maximum payload rating. As this 797 passes by, it's payload display reads 385 tons. *ECO*

This is what the driver of a 797 sees out of the right side of the cab as the truck is being loaded. Looking at the right side mirror, which is nearly 30 feet away from the operator, one can just make out the activity to the rear of the hauler. In areas close to the truck, the operator's field of vision is severely limited. That is why Caterpillar offers an optional camera system. Cameras mounted in the front keep an eye on who might be approaching the front access ladder at ground level, and rear-mounted units aid in backing the hauler into position under the loading shovel. *ECO*

in June 1999. The next batch of 797s found their way up to Fort McMurray in Northern Alberta, for use at Syncrude Canada Ltd.'s oil sand mining operations. The first trucks delivered in 1999 were assigned to Syncrude's North Mine, with additional units delivered in 2000 for use at the company's Aurora Mine near Mildred Lake. Another oil sand mining operation, Suncor, also took delivery of a fleet of the giant Cats. Other mining locations for the 797 in 2000 included the Powder River Basin in Wyoming, and selected deep pit copper mining operations in South America.

At the end of 2000, two special 797 HAA (High Altitude Arrangement) haulers were shipped to P.T. Freeport Indonesia's Grasberg gold and copper mine located in Indonesia at an altitude of 13,120 feet. The special drivetrain setup is necessary to cope with the thin atmospheric conditions found at these extreme working altitudes. Freeport received its first 797 in November 2000, followed by a second unit in December. Full production for the 797 for the entire world market got under way in 2001.

In operation, the 797 rides and handles much like the company's other large mechanical-drive off-highway trucks. A loaded 797 has no problem pulling away from its loading area, even in less than ideal conditions. The engine's 3,211 flywheel horsepower moves the truck out smartly and gets it up to speed quickly. For long uphill climbs out of some pit locations, the engine's broad torque curve ensures that the power band is just were you need it. The ride is exceptionally smooth, assisted in part by the operator's air-ride seat. A second air-ride seat is also provided for another passenger. During hard braking, the nose dives only slightly. The operator's ROPS cab has all the comforts of a well-equipped car, such as power windows, tilt and telescoping steering column, air conditioning, stereo, and even a cup holder. On the dashboard is Caterpillar's Vital Information Management System (VIMS) display and keypad, which provides precise machine status information to the operator. The cab is resiliently mounted to dampen noise and vibration. The cab offers good lines of sight for the operator, and the clean design of the upper deck gives a clear view to the right of the truck. This is important since the right-side rearview mirror is almost 30 feet away from where the operator is sitting. A person who has poor depth of field vision maybe should not get behind the wheel of the 797. An optional camera system is also available to allow the operator to see directly in front of and to the rear of the hauler.

Out in the field, the 797 has performed as anticipated. Some problems have cropped up during testing, but this was expected by Caterpillar engineers. This is why the 797 program initially started out with 20 or so field-follow units. Problems could be addressed and corrected, and implemented immediately to trucks currently in the pipeline. Improvements have included a reinforced

On the haul road, the 797's cab has all of the comforts and operating information a person could want. With Caterpillar's Vital Information Management System (VIMS), the operator has a real-time display of all of the truck's vital systems. The 797 utilizes a seven-speed, electronically controlled automatic power shift transmission. In top gear, on level ground, a fully loaded 797 can attain a top speed of 40 miles per hour. *ECO*

frame, larger rear differential coolers, and a new engine module frame design. All of these improvements have added to the 797's overall operating weight. Upon introduction, the truck weighed in at 1,230,000 pounds. As of 2001, the big hauler tips the scales at 1,345,000 pounds, a weight gain of almost 58 tons.

One area of confusion regarding the 797 is its rated payload capacity. The hauler is rated at a nominal, or normal, capacity of 360-plus tons. It's that "plus" that is confusing. There are a host of mine-specific conditions that dictate how a truck the likes of a 797 is set up. Haul road distance, grades out of the pit, weight of the material, and geographic and weather conditions all play a part in determining the truck's ultimate working payload.

Another factor that limits the hauler's tonnage capacity is its tires. Overloading of the 797 has led to some premature tire wear. As large as the Michelin 55/80R63 tires are, the heat buildup caused by excessive loading, high travel speeds, and long load haul distances, is tremendous. This eventually leads to the breakdown of the side walls and carcass of the tire. Designing a tire for a truck like the 797 is no walk in the park. The tire has to perform in extremely hot conditions, such as those found in Arizona, or in the subfreezing temperatures of Northern Alberta, which can reach -30 to -40 degrees Fahrenheit. Because of the type of material being mined by Syncrude, they actually get the highest productivity out of their truck fleet in the winter months. The haul roads are frozen solid, so rolling resistance is at a minimum, and the freezing temperatures keep the tires cool. During these times, the 797 often finds itself with 400 tons on its back. But during the spring and early summer, when the ground is very wet and spongy, the rolling resistance is at its highest. This, along with warmer atmospheric conditions, dictates that loads be restricted to 360 to 380 tons.

While mines adapted to tire limitations, a better solution was forthcoming. In October 2000, at the MINExpo exhibition in Las Vegas, Caterpillar showcased a 797 equipped with newly designed Michelin 58/80R63 tires, the largest ever molded for off-highway hauler use. Mounted on 44-inch-wide rims (the original rim width was 40 inches), with a 58-inch overall tread width, the new tire provided performance and durability better suited to the truck's capabilities. Starting in 2001, this became the factory-recommended tire. Eventually, all 797s were retrofitted with this larger tire and wheel package.

Caterpillar offers the 797 with a standard flat floor dump body, or special Mine Specific Design (MSD) bodies. These dump bodies can be fabricated to meet a host of material hauling needs. Lighter weight bodies can be specified for less abrasive material. Weight subtracted from the dump body's design adds directly to the payload capacity of the truck. Other designs have different floor configurations. It all depends on the material

To get the most out of a 797, a mining operation needs to employ the biggest mining shovels around. The capacity of this shovel's bucket is 100 tons per dipper full. The shovel operator will place three full loads into the back of the truck, and hold off just a little on the fourth, so the truck is not overloaded. This is quite easy to do with a bucket of this size. Overloading the truck, even as one as large as the 797, can lead to shorter component life and severe tire wear. *ECO*

that will eventually be dumped from the back of the truck. In Wyoming, 797 haulers equipped with specially designed coal bodies, fabricated by WOTCO of Casper, Wyoming, regularly average 380 to 386 tons per trip. Their voluminous 588-cubic-yard bodies are simply huge, especially when compared to the standard 290-cubic-yard box. This capacity is possible since coal is lighter than, say, overburden.

The Competition

The Caterpillar 797 is without question the world's largest mechanical-drive mining truck. But there are a few other off-highway trucks currently on the market that classify as ultra-haulers, and have payload capacities very similar to that of the big Cat. Built by such manufacturers as Komatsu, Terex Unit Rig, and Liebherr, all have one thing in common—they are designed around AC diesel-electric drive systems.

As mentioned above, the Komatsu 930E was the first of the two-axle, ultrahauler class of mining trucks equipped with AC drive. As of 2001, the company offered two model types of this well-received truck, the 930E-2, and the 930E-2SE. Both haulers are rated as 320-ton-capacity machines, and ride on 53/80R63 tires on 63-inch-diameter rims. The biggest difference between the

A fully loaded 797 pulls itself out from under the loading shovel, as it starts to make its way out of the pit area. This version of the 797 is equipped with a lightweight MSD body and improved 58-series tires, which increase the truck's overall width to almost 31 feet at the rear. The unit also features a slightly revised upper-deck engine air-cleaner arrangement, which does away with the previous design that utilized a protective canopy. *ECO*

two model types is the engines. The standard truck is equipped with a Cummins QSK60 diesel engine rated at 2,550 flywheel horsepower (2,700 gross). The 930E-2SE is featured with a newly designed Komatsu SSDA18V170 diesel engine, developed in cooperation with Cummins, which refers to it as the QSK78. This massive engine is an 18-cylinder configuration with 12 turbochargers, making it suitable for high-altitude operations. The big V-18 is rated at 3,429 flywheel horsepower (3,500 gross). With this much power on hand, one would assume that the capacity would also rise. But this is not the case. The 930E-2SE is designed for increased productivity through speed out of the pit. This high-powered Komatsu was first seen at the MINExpo exhibition in October 2000, and has been going through field evaluation testing throughout most of 2001.

Another competitor to the Caterpillar 797 is the Terex Unit Rig MT-5500. This 360-ton-capacity hauler was first unveiled at a special ceremony at the company's Tulsa, Oklahoma, assembly plant in April 2000. Like the Komatsu 930E-2/2SE models, the MT-5500 utilizes an AC diesel-electric drivetrain. The MT-5500 is rated at 2,478 flywheel horsepower (2,700 gross). The first MT-5500 was shipped to a coal mining operation in the Powder River Basin region of Wyoming, with the next group of trucks destined for South America.

The last challenger is the Liebherr T-282. Built in 1998, the 360-plus-ton-capacity truck was officially unveiled in October of that year, only about two weeks after the unveiling of the Cat 797. The original T-282 featured a Detroit Diesel/MTU 16V4000 diesel engine, rated at 2,750 gross horsepower. The Liebherr utilizes 63-inch rims, 55/80R63 tires, and AC diesel-electric drive. At the MINExpo 2000 show, Liebherr showcased an even more powerful version of the T-282, equipped with a state-of-the-art V-18

Cummins QSK78 diesel engine. This engine, the same unit found in the Komatsu 930E-2SE, was rated at 3,500 gross horsepower. The truck also featured a special lightweight body, which, when combined with the massive power of the V-18, raised the maximum capacity of the T-282 to 400 tons. Liebherr has 360-plus-ton-payload-capacity trucks currently working in Wyoming, and in Northern Alberta. As of 2001, the 400-ton-capacity model was still undergoing further developmental testing.

All of these trucks are huge, with dimensions similar to the 797. The Komatsu measures 27 feet, 4 inches to the outside of its rear tires; the Terex model measures 30 feet, 2 inches; and the Liebherr comes in at 28 feet, 7 inches. These are all very close to the Caterpillar 797's 31-foot width. The overall loaded working weight of these trucks is also very similar. The Komatsu weighs in at 1,100,000 pounds, the Terex comes in at 1,198,000 pounds, and the Liebherr tips the scales at 1,248,000 pounds. The big Cat beats them all with a gross weight of 1,345,000 pounds. Though all of these weights can fluctuate, depending on the type of dump body installed, the Cat truck is clearly the heaviest of the group. The 797's robust frame and mechanical drivetrain do add a bit of weight to the truck's bottom line. But it is capable of handling it with the aid of its powerful V-24 and new 58-inch-wide tires.

All of the ultrahaulers are technical marvels. Each has its strong and weak points. But in the end, it is the customer that makes the final decision as to what truck will live or die in the marketplace. The 320-ton Komatsu has a three-year jump on the 797. But sales have been strong with the 797. At the time of this writing, 797 haulers were in operation in Canada, Chile, and Indonesia, as well as Arizona, Utah, and Wyoming.

Though the tonnage display reads 446 tons on this 797 as it gets up to speed, things are not always what they appear to be. While the truck is being loaded, the shock of the load as it hits the dump body is also being recorded. Once the 797 is under way, it will reweigh itself a few seconds after it upshifts into second gear. The actual load on this truck was 386 tons. *ECO*

The largest motor grader offered by Caterpillar today is its model 24H. Released for sale in 1996, it is presently the world's largest production motor grader offered by any equipment manufacturer. *ECO*

CHAPTER THREE

The Biggest Blade in Town

24H MOTOR GRADER

During the late 1920s, the Caterpillar Tractor Company made a strategic business decision to expand its presence in the marketplace. Up to that time, Caterpillar was known primarily as a supplier of track-type tractors for agricultural use. Going forward, it would now be linked with equipment targeted at road construction and maintenance as well. In August 1928, Caterpillar purchased the Russell Grading Manufacturing Co. of Minneapolis, Minnesota. Considered one of the leading companies in the building of road maintenance equipment, its products complemented Caterpillar's tractors perfectly.

The Russell company was founded in 1903, by Richard Russell and C.K. Stockland, in Stephen, Minnesota. The company's first major product was a horse-drawn elevating type of grader, with a gas engine-driven conveyer. With the success of this early design, the company moved to larger facilities in Minneapolis in 1906. The larger manufacturing plant allowed Russell to expand its road grader offerings, with such products as blade graders, drag and wheel scrapers,

The first Auto Patrol undergoes field-testing in 1931, the year Caterpillar introduced it as the industry's first true motor grader. What made the Auto Patrol unique was that it was designed from the ground up as a motor grader. Previous designs utilized grader attachments combined with existing tractor designs. *Caterpillar, Inc.*

and plows. The first of these implements, a small two-horse road maintenance grader, was introduced in 1908. This was followed the same year by the first eight-horse-pulled Simplex Road Machine, and then the Traction Special, a grader specifically built for use with a tractor. Other popular Russell blade graders would soon emerge, carrying such creative names as the Standard (1912), the Mogul (1913), the Super-Mogul (1922), the Reliance (1915), the Super-Reliance (1923), the Special (1912), and the Super-Special (1924). All were designed to be pulled by a tractor of some type.

In 1919, the humble little Russell road grader evolved into something much more, after the company added an engine to the equation. This experimental machine was in fact the world's first self-propelled grader. The little invention consisted of a single-axle, two-wheeled Allis-Chalmers tractor in the front, with a grader unit attached at the rear. Production would officially commence in 1920 as the Russell Motor Hi-Way Patrol No. 1. Interest in the marketplace was high, but when Allis-Chalmers ran into a patent dispute with that particular tractor model, it had to be withdrawn from the market. The Russell self-propelled grader would have to be put on hold, until another suitable tractor power source could be found.

In 1925, Russell released a new model of the self-propelled grader design called the Motor Patrol No. 2. This design utilized a Fordson tractor at the rear, with the grader attachment moved to the front. Unfortunately, Fordson showed little interest in further developing the project, so the model was eventually dropped after only a couple years of production. But the design of the grader with the tractor in the rear was sound. Other Russell designs in this configuration included the Motor Patrol No. 3 in late 1925, based on McCormick-Deering's Model 10-20. Caterpillar would enter the picture in late 1926, when it supplied its

Two-Ton crawler tractor for the Russell Motor Patrol No. 4. In 1927, Motor Patrol No. 5 was offered with a Cletrac K-20 crawler tractor, and in 1928, the Motor Patrol No. 6 was released based on the Caterpillar Twenty.

After purchasing Russell in 1928, Caterpillar immediately went to work revamping the grader product lines to fit its own track-type tractor offerings. The company quickly sold off product lines that did not have a direct bearing on the sale of Caterpillar tractors. Any Motor Patrol model that did not utilize a Caterpillar tractor was terminated. The product lines that were retained consisted of self-propelled, tractor-pulled, and elevating graders.

The new Caterpillar self-propelled graders included the Motor Patrol Models No. 10, with a Cat Ten tractor (1929); the No. 15, with the Cat Fifteen (1929); and the No. 20, formerly known as a Motor Patrol No. 6, with the Cat Twenty (1928). The last of this type of design to be released by Caterpillar was the Motor Patrol Model No. 28 (1933), which utilized a Cat Twenty-Eight tractor. All of these designs relied on a separate tractor model and grader attachment. Caterpillar engineers were quick to realize that it would make far better sense to combine the power tractor unit and the grading mechanism into one cohesive design. But this new design would be centered around a tire and wheel drivetrain layout, without the use of any type of crawler track assembly. So was born the "Auto Patrol."

Built in 1931, the Caterpillar Auto Patrol was the earthmoving industry's first true rubber-tired, self-propelled production motor grader. Unlike the earlier Motor Patrol model lines, on which the grader was simply a front-end attachment mounted on an existing crawler unit, the Auto Patrol had its own engine placed high and in the rear of the machine. This kept the engine in a

The Caterpillar No. 12 was without a doubt one of the greatest motor grader models ever offered by the company. Originally released in 1938, it set the standards for its size and power class for decades to come. Originally released with diesel or gasoline power, the gas version would end production by 1942. This Diesel No. 12 from the mid-1950s shows just what kind of incredible working situations this motor grader could handle. *ECO Collection*

Another milestone in Caterpillar motor grader design was its famous model No. 16. Introduced in early 1963, it was the most powerful grader design ever offered by the company at the time. With 225 flywheel horsepower and a 14-foot moldboard (16-foot optional), the No. 16 was built to take on the big highway-building contracts of the day. *Caterpillar, Inc.*

cleaner work environment, improved operator visibility, and increased traction on the drive-axle. So solid was the design concept, it would form the basis for all motor graders yet to come in the industry. Even today, the basic design layout of the first Auto Patrol is still with us.

At first, Caterpillar's new grader was simply referred to as an Auto Patrol. But by the end of 1931, it received the model No. 9 designation. In early 1932, a slightly lighter weight version of the motor grader was introduced called the No. 7. Production on these first two Auto Patrols would end in early 1933.

Caterpillar was quick to add new and improved Auto Patrols starting in late 1932 with the No. 11. This was followed by the No. 10 in 1933. Both were gasoline-powered motor graders, with a single rear drive axle. Starting in 1934, Caterpillar introduced the Diesel Auto Patrol. This was also the first year that the company offered tandem-drive rear axles on all of its Auto Patrol models, which helped relieve the bouncing or "loping" ride that the single rear-axle graders commonly suffered from. Starting in 1937, the Diesel Auto Patrol became the Diesel No. 11. Also added to the line-up was a Diesel No. 10. Both models were available in single- or tandem-drive configurations, just as in the gasoline-powered versions. Leaning front wheels were offered for the first time as an option in 1937—an important

engineering step in the company's young motor grader line-up.

In mid-1938 Caterpillar released one of the most significant motor grader designs of all time, the model No. 12. Key to the No. 12's success was its triple-box section main frame, which was far more rigid and stronger than that of the old twin-beam designs found in previous Auto Patrols. Leaning front wheels became standard fare on the No. 12, making the grader's turns shorter, and counteracting side-draft caused by grading forces at the blade. The original model, which was referred to as a No. 12 Auto Patrol in its first year only, could be ordered with either a diesel or gasoline engine. But by 1942, the gasoline engine option had been dropped from the line-up. A tandem rear axle drive layout was the only choice.

At the time the No. 12 motor grader was introduced, the No. 11 and No. 10 Auto Patrols were still being offered. But by mid-1939, Caterpillar replaced both model lines with fresh designs identified as the No. 112 and No. 212. The older Auto Patrol models were still built until their parts inventories were exhausted in early 1940. The new models were basically smaller versions of the No. 12, with the No. 112 being the larger of the two. Each was initially offered with either diesel or gas engine, as well as single- and tandem-axle rear-drive layout. By 1947, gasoline power was dropped.

The legacy of the original No. 16 motor grader lives on today in the form of the model 16H. Introduced in late 1994 as the replacement for the 16G, this 275-flywheel-horsepower grader is the second-largest model produced by Caterpillar at the present time. *Caterpillar, Inc.*

CATERPILLAR
24H

Caterpillar 24H motor graders are built to withstand the hottest operating working environments, or in this case, the coldest. This 24H is hard at work during the wintertime in British Columbia. This unit is also equipped with higher-mounted rear fenders, which allow tire chains to be installed for ultrasevere winter haul road work. *ECO*

Primarily designed to maintain haul roads within today's large open-pit mining operations, a single 24H can do the work of two 16H units. This view of a 24H in its element shows that a typical large operation has miles and miles of haul roads that need to be taken care of 24 hours a day, 7 days a week. *ECO*

During the 1950s and early 1960s, Caterpillar designed and introduced a number of fine motor graders. Some of these included the No. 12E and the No. 112E in 1959, the No. 112F in 1960, the No. 120 in 1964, and the No. 12F in 1965. Most of these offerings were in the 100- to 115-fly wheel-horsepower class. The company significantly increased power output with the No. 14B from early 1959. This model was the company's first motor grader to reach 150 flywheel horsepower. Other models of the No. 14 grader included the "C" in late 1959, the "D" in 1961, and the "E" in 1965. All of these motor graders were considered highly productive road-building tools. But what Caterpillar lacked was a "big" grader, one with the power and blade capacity to tackle the largest earthmoving contracts. That would change in early 1963 with the introduction of the No. 16.

The No. 16 motor grader was designed and built for contractors looking for a machine to help them meet the demands of large highway contracts. The No. 16, which evolved from a Caterpillar experimental motor grader identified as the No. 865 from the late 1950s, was nearly twice the size of a No. 12 and half again as large as the No. 14 series. The No. 16 was powered by the ultra-reliable Cat D343 diesel engine, rated at 225 flywheel horsepower. This was mated to a torque-divider powershift transmission, which allowed the operator to select from three speed ranges and three types of drive to match the grader to any kind of working condition. The transmission gave the unit nine forward speed selections in all. Tandem rear drives afforded the grader exceptional traction in the worst conditions. Other key performance features included oil-cooled disc brakes on the rear drive wheels, a rugged triple-box steel frame design, and power-boosted mechanical controls, a first for a Caterpillar motor grader. The standard moldboard on the No. 16 was 14 feet in length and 31 inches in

height, with optional 15- and 16-foot blades available. Overall working weight was about 46,500 pounds. A rear-mounted ripper-scarifier was also part of the options list, which added an additional 4,700 pounds to the operating weight.

The No. 16 proved to be a stellar performer. Its overall speed, productivity, and reliability were unmatched in the industry. Though there were a few machines on the market that were actually more powerful than the No. 16, none could match its overall performance and balance in the field.

Caterpillar kept the No. 16 grader program supplied with a steady flow of improvements and upgrades that were constantly sending the competition back to the drawing boards. In 1973, Caterpillar took the wraps off of the newly redesigned 16G motor grader. This new model series incorporated numerous design and performance upgrades. Of most importance, the 16G now featured an articulated frame, which pivoted right behind the operator's cab. This layout allowed the operator to choose from three main frame positions—straight, articulated, or crab, depending on the job requirements. The blade controls were now completely hydraulic, making blade positioning faster and more precise. A new, rear-mounted 250-flywheel-horsepower, Cat 3406 diesel engine provided the power, mated to a single-lever powershift transmission with eight forward speeds. The 16-foot moldboard became the standard blade in the model series, with the 14-foot unit optional. Equipped with a ROPS cab and rear-mounted ripper, the 16G was a 58,900-pound blade with an attitude. It was the company's finest and most modern motor grader to date. The earthmoving industry simply loved it.

The 16G remained essentially unchanged during its production run, but did get a power boost in 1985 to 275 flywheel horsepower. Over the years the company also made numerous system upgrades, such as an Electronic Monitoring System (EMS), which kept an eye on the grader's

The Caterpillar 24H is currently king of the motor graders. Measuring almost 52 feet in length, and weighing in at 136,611 pounds, the 24H is the ultimate motor grader for mining haul road maintenance. *ECO*

Here, the front frame assembly of a 24H makes its way down the large motor grader assembly line at the Decatur facilities. The frame is a flanged box-section design, which provides superior frame rigidity. *ECO*

vital functions so the operator could concentrate fully on the job at hand. The final version of the 16G weighed in at 60,150 pounds in full operating trim.

In late 1994, the 16G evolved into the 16H. The new "H" upgrade in fact covered the entire product line of all Caterpillar motor grader models offered. Upgrades to the 16H were subtle, but noteworthy. These included an improved high-torque Cat 3406C diesel engine, which offered improved overall lugging performance. Though the torque curve was up, the horsepower output remained constant at 275. Other features included a new load-sensing hydraulic system; diagnostic capability, allowing fast servicing of the starting-charging system; and an improved operator's cab with larger windows, lower interior sound levels, repositioned gauges, and improved low-effort controls. Overall operating weight with rear-mounted ripper totaled 58,858 pounds in standard form. All of these enhancements made the 16H an even more productive machine—something customers may have found hard to believe. The 16-series motor graders have sold in the thousands and are considered some of the finest machines of their type ever built by any manufacturer. Their place in earthmoving history is secure.

The 24H

In the late 1980s, as mining operations began to employ bigger fleets of large-capacity off-highway haulers, their haul road infrastructures were becoming more and more of a concern. Good haul roads are essential to mining operations. As the haul trucks increased in capacity, their overall size and weight also increased. To keep these massive truck fleets moving, mine roads needed to be wider and better constructed. Though mining operations could make do with motor graders of the 16-series size, they had to use a small fleet of

them, each requiring its own operator, in multiple shifts. The cost of maintaining large fleets of graders was increasing the costs to the customer substantially. Caterpillar soon realized that what its mining customers needed was not more motor graders, but larger and more productive units that could take the place of multiple machines. Caterpillar management soon approved funds for the development of a bigger motor grader, far larger than its 16G model, which was current at that time. With funds approved, Cat engineers started to work on what would eventually become the 24H.

Like its mining equipment cousins, the 994 loader and the 793 hauler, the 24H started its life at Cat's Mining Vehicle Center (MVC). During the early design phases of the grader's development in the early 1990s, it was initially referred to internally as the 18H project. This reflected the unit's planned 18-foot-wide blade. But as the grader advanced in design, research showed that an even larger unit would be needed to do the job that Cat had in mind. Haulers in excess of 300-tons capacity were soon to be introduced to the mining industry. To support these haulers, Caterpillar engineers calculated that a grader with a 24-foot blade would soon be needed. By 1994, the overall design criteria for the 24H were locked in stone as the engineering and design team started to construct the first prototype.

The 24H was designed for mining operations that utilize 150-ton-capacity and larger haul trucks. Its principle task is to build and maintain haul roads. Unlike the 16H grader, which was designed to perform a broad range of grading and ditching functions, the 24H was built as the ultimate management tool for the mining industry in which larger haul roads were a significant concern. The 24H can do the work of two 16H graders. Jobs that formerly required four 16H graders would need only two 24H units. Even though the

At the end of the assembly line, this 24H undergoes a series of pressure checks to make sure all hydraulic systems are up to specifications. The tires on this unit are "shop-tires" and are for testing at the plant only. A new set will ship from their point of manufacture directly to the customers' mine site. *ECO*

Once all of the main powertrain components check out, the 24H will make its way outside of the main plant for preliminary running and braking tests. The grader's 24-foot moldboard will not be installed until the machine reaches its final assembly destination. *ECO*

24H costs considerably more than the 16H, it offers a lower operating cost per haul road mile because of decreased overall maintenance and crew requirements. Bigger is better when it comes to maintaining mining haul roads.

The 24H achieves its impressive production output with power, weight, and one mighty big blade. The grader is powered by a rear-mounted 12-cylinder Cat 3412E HEUI, turbocharged diesel engine, rated at 500 flywheel horsepower (540 gross). This power plant is mated to an electronic-shifting Cat powershift transmission with six forward speeds, giving the grader a top traveling speed of 23.4 miles per hour. Though some Caterpillar grader models offer six-wheel-drive capability, only the rear tandems are driven on the 24H. The frame of the grader is articulated just behind the cab, giving the unit enhanced maneuverability in tight working situations. An optional rear-mounted ripper with up to seven ripping shanks is also available. Most 24H graders ship with the ripper option. Overall working weight is 136,611 pounds, more than twice that of the 16H. With this type of power and weight, the 24H can apply greater cutting forces to its blade, enabling it to take on the most difficult road surfaces.

The business end of the 24H is its standard 24-foot-long moldboard. This blade, which measures 42 inches in height, has the ability to move almost 2 1/2 times more material than the 16H.

A 24H reaches its top running speed of 23.4 miles per hour on the special paved oval test track behind the Decatur assembly plant. Machines are put through their paces on the track to check for possible drivetrain and suspension problems. Products run on the test track include scrapers, haul trucks, and all of the motor graders. The heavy mining haulers, such as the 785C, 789C, and 797, are not tested in this fashion, since they ship in subassemblies and are too heavy to be supported by the track itself. *ECO*

In fact, the 24H produces 40 percent greater down pressure on the blade than the 16H. And with the 24H's wider blade, the grader needs fewer passes to do the same job. Caterpillar also offers a wide selection of condition-specific cutting edges for the blade, allowing customers to match cutting edges to the abrasion characteristics and penetration requirements of their haul road materials. A blade the size of the 24H is essential for economical road building within mine sites. As a general rule of thumb, two-way haul roads need to be three to four times the width of the trucks being used. And with mining haulers such as the 797 measuring 30 feet across, one starts to get a clearer picture of just how much roadway needs to be built and serviced, 24 hours a day, 7 days a week.

Caterpillar delivered its first preproduction 24H in late 1995 to the Eagle Butte coal mine, located just north of Gillette, Wyoming. Other units were placed throughout North America for field follow-through testing. The 24H was

24H
622

The front wheels of the 24H have the ability to "lean," which keeps the grader traveling in a straight line while the blade is engaged. The rear tandems are driven, giving the unit four-wheel drive. The front wheels, on the other hand, are not under power. The 24H steers by means of its front wheels and an articulated frame, which pivots just behind the operator's cab. *ECO*

The 24H has walkways on either side of the rear engine bay, making service quick and easy. Opening the large removable side doors gives full access to the 500-flywheel-horsepower Cat 3412E V-12 turbocharged diesel engine. *ECO*

officially released for worldwide availability around May 1996. Though originally assembled at Caterpillar's Joliet, Illinois, plant, production has since been shifted to the Decatur facilities. This was done so the 24H's production and marketing could be integrated with that of the other large mining equipment product lines also being built at the plant.

The Competition

Caterpillar's two largest motor graders are in an enviable place in the marketplace today. Unlike other Caterpillar mining equipment lines which face numerous competitive challenges in the field, the large motor grader lines escape nearly unchallenged. The 16H faces only two rival machines that compete directly for its market share. One is produced by Volvo, and the other by Komatsu. The Volvo 780A VHP is a 235-flywheel-horsepower, 47,360-pound motor grader, equipped with a 16-foot moldboard. This unit, which was formerly identified as a Champion product, is a bit smaller than the 16H, but nonetheless utilizes the same length blade. The model from Komatsu is the GD825A-2. This motor grader is a 280-flywheel-horsepower, 58,250-pound design equipped with a standard 16-foot, 2-inch moldboard. It is the only grader currently in production that can match the Caterpillar offering, at least on paper, specification for specification. Though Orenstein & Koppel of Germany once fielded a large grader identified as the G350 in late 1979, with 360 gross horsepower and a 19-foot blade, it went out of production in the early 1990s, with only 34 units built. In the end, all that really counts is sales, and here the 16H reigns supreme.

As for the Caterpillar 24H, it really has no direct rival in the marketplace. In fact, the 24H is currently the largest production motor grader available to the mining industry. True, there have

been graders built in the past that were far larger than the 24H, but all were considered limited production machines. Today, all of these designs are out of production. Grader models from the past include the experimental RayGo Giant from 1969. This model was a twin-engined, 636-gross-horsepower design, equipped with a hefty 8,000-pound, 20-foot blade. Operating weight of the Giant was 105,200 pounds.

A far larger machine was the Champion 100T from 1978. This ultralarge grader, originally identified as the 80T in 1975, was rated at 700 gross horsepower and weighed in at 202,000 pounds. Its moldboard was 24 feet in length, with a 56-inch height. It quietly went out of production in the early 1990s under the Dom-Ex brand name.

Yet the largest of all past motor graders was the Italian-designed experimental ACCO Grader from the early 1980s. This massive machine was powered by two engines providing 1,700 gross horsepower. All six wheels were driven, with two tires mounted to each wheel assembly, for 12 tires in all. The moldboard measured 33 feet across in standard form. Operating weight was right around 200 tons, which is only about 22 tons less than the Cat 24H, RayGo Giant, and Champion 100T weights combined! The ACCO design was big, and then some. But its size could not save it in the marketplace. In fact, only two were ever produced, one prototype and one pre-production machine.

All of these motor graders from the past promised big production and high productivity for the mining customers that considered them. But all would fail in their designed mission.

Today, if a mining operation needs the biggest blade in town, it need look no further than the local Caterpillar dealer, home of the world's largest *production* motor grader, the 24H.

The 24H comes standard with a rear-mounted three-shank D6-size ripper. The ripper can accommodate up to seven shanks, depending on the type of material and working conditions. *ECO*

The 994 wheel loader is a massive, high-production loading tool designed for use in the world's largest mining operations. Measuring 55 feet, 9 inches in length when equipped with the high-lift option and 53.5-series tires, the 994 weighed in at 401,740 pounds. *ECO*

CHAPTER FOUR

The Big Wheel Loaders and Dozers

994 WHEEL LOADER

In the 1950s, Caterpillar added many new product lines aimed specifically at the construction side of its business interests. Rubber-tired, self-propelled scrapers were introduced, along with track-type loaders, in the early 1950s. By 1959, the company released its first rubber-tired front-end loader, the model 944A Traxcavator. The loader was first unveiled to the company's board of directors as the T101, at a special demonstration at its Phoenix Proving Ground in February 1956. Though not a large machine, its standard bucket capacity of 2 cubic yards put it right in contention with a host of similar-sized machines from other construction equipment manufacturers. The unit's four-wheel-drive drivetrain, rigid frame, and rear steering wheels were design features well accepted in the industry. Caterpillar was looking for the perfect sized wheel loader design to get its latest product line off to a good sales start, and it found it in the 944A.

But the model 944A was just the tip of the iceberg as far as Caterpillar was concerned. In 1960, the company introduced two more rigid frame wheel loaders to keep the 944A company. The model 922A was slightly smaller, carrying a 1 1/4-cubic-yard bucket. The other model, the 966A, was larger than the 944A. It was equipped with a 2 3/4-cubic-yard bucket. All three of these loaders were sized for general contracting work. Big quarry and mining wheel loaders they were not. But Caterpillar was taking a cautious approach to this new market. The limited number of model types fielded in the first few years allowed the company time to gain market share, but also to perfect a new type of loader design that was starting to gain increased attention in the industry. This design did not steer by its rear wheels, but by a hinged, or articulated steering frame. The articulated-steering wheel loader was far more maneuverable than a rigid type, and offered superior productivity gains in all working conditions. Though Caterpillar's rigid frame loaders were well received in the industry, these three

In late 1965, Caterpillar started testing a new wheel loader design far larger than its current 988 series. Identified as the 992X1, this mining-sized front-end loader was powered by a Cat D346 V-8 diesel engine and was equipped with an 8.5-cubic-yard bucket. The cab on this prototype was an after market design supplied by Industrial. *Larry Clancy Collection*

In 1977, Caterpillar introduced what is still considered one of the finest and best-selling large wheel loaders ever designed, the 992C. Its 690-flywheel-horsepower Cat 3412 V-12 diesel engine, and its big 13.5-cubic-yard bucket made it a favorite among operators worldwide. *Caterpillar, Inc.*

The popular Caterpillar 992C was replaced in 1992 by the improved 992D. Features of the new loader included the use of the joystick control STIC System for steering and an increase of payload capacity to 14 cubic yards. The 992D also featured an access platform over the rear left tire. *Caterpillar, Inc.*

models would seem downright antiquated when compared to new models the company had up its sleeve. These models were the 966B and the 988.

Both the 988 and 966B Traxcavator articulated-steering wheel loaders were introduced to Caterpillar officials at a special Phoenix Proving Ground demonstration in February 1962. But it would be months before either loader was ready for commercial release. Finally, in January 1963, the 988 was officially sanctioned for sale, followed by the smaller 966B in June of that year. The 988 was a 300-flywheel-horsepower, 58,500-pound loader, with a bucket payload capacity ranging from 5 to 6 1/2 cubic yards. The 966B came with 150 flywheel horsepower, a 31,000-pound operating weight, and a bucket capacity of 2 1/2 to 5 cubic yards.

The new 988 was almost twice as large as the 966B. It is no coincidence that the 988 was introduced in the same month that the newly designed 769 off-highway truck reached full production status. The size of the 988 made it the perfect match for the 769, along with the rest of the 35-ton-capacity haulers in use at the time of its introduction. Caterpillar marketing often promoted the virtues of the articulated 988 in conjunction with the 769. This was further reinforced with many of the company brochures featuring 988 loaders loading 769 trucks. These complementary vehicles gave the company's dealerships an edge in marketing Caterpillar products to large quarry operations. The arrangement worked well for the dealerships, the customers and the company.

During the 1960s, the Caterpillar 988 received steady improvements and power increases to keep it in step with the marketplace. By 1968, the loader carried a 325-flywheel-horsepower rating and weighed in at 68,000 pounds. The largest bucket specified was a general purpose 6 1/2-cubic-yard unit. The 988 was a fairly large machine for its day, but the mining industry had an appetite for ever larger and more productive equipment—an appetite to which other manufacturers were already responding. Though the 988

Caterpillar replaced the 992D loader with a completely new design in the form of the 992G. Introduced at the 1996 MINExpo show, the 992G featured the company's first use of a unique one-piece, cast-steel box-section front lift arm. Though this type of mono-boom configuration had been tried by other manufacturers over the years, it was only utilized on smaller machines and never on anything on the scale of the 992G. *Urs Peyer*

could be found in mining operations the world over, its real sales success was in the quarry and aggregate industries. To compete in the wheel loader classes designed for large-scale mining operations, Caterpillar would need a bigger and more powerful machine.

As the 988 loader began to find its feet in the marketplace, Caterpillar engineers set their sights on this much larger machine. The new loader would be designed to match haul trucks in the 50- to 90-ton-payload-capacity class. There were already many trucks in the marketplace that would be a perfect match for this loader. If designed correctly, it could guarantee solid sales returns for the company for years to come.

The first pilot machine, referred to as 992X1, emerged in prototype form in late 1965. At this point in the design, the loader was powered by an eight-cylinder, Cat D346 diesel, rated at 500 flywheel horsepower. Capacity was 8 1/2 cubic yards. It was also equipped with an after market industrial operator's cab. In January 1966, the 992X1 was shipped to Clarkson Construction Co. in St. Louis, Missouri, to start its field-testing trials. After months of design evaluations, a second prototype, 992X2, was completed in December 1966. This unit was powered by a more powerful Cat D348, V-12 diesel engine, rated at 550 flywheel horsepower. The 992X2 was rated as a 10-cubic-yard machine, and rode on tires that were larger than those fitted to the earlier 992X1. It was equipped with a Cat-designed operator's cab that was pressurized and air-conditioned, along with improved ladders and steps. Over the next few months of testing, this design proved that it had what it took to become a production reality. In October 1968, the first production 992 loader officially went into service at the Duval Mines in Twin Buttes, located south of Tucson, Arizona. It would be the first of many 992 machines, a model that would eventually become the world's best-selling wheel loader in its size class.

The 992 looked much like the 988, but was larger all the way around. With its standard

10-cubic-yard rock bucket, the loader had a maximum payload capacity of 30,000 pounds. Its overall operating weight with this bucket was 120,500 pounds—almost twice the weight of the 988. With its 550-flywheel-horsepower, turbocharged V-12; planetary full power shift automatic three-speed transmission; and articulated frame, the 992 proved to be one tough loader out in the field. But this was just the start of the 992 model line's evolution.

In 1973, Caterpillar introduced an updated B series of the 992. The 992B contained numerous mechanical changes aimed at increasing the machine's reliability and productivity. The power plant and the horsepower output were unchanged from the 992 model. New for this model were a pin-on ROPS cab, which protected the operator in the event of a rollover, and completely sealed oil-cooled disc brakes. Though the loader's capacity remained unchanged from the previous model, the operating weight was now up to 135,300 pounds. Like its predecessor, the 992B kept raking in the sales.

In the late 1960s and early 1970s, some heavy equipment manufacturers had brief love affairs with turbine-powered mining equipment. Caterpillar too looked into the benefits of the high power-to-weight ratios of the turbine engine layout with an experimental 992 Gas Turbine Loader from August 1973. This prototype resembled the 992B in its front end and ROPS cab, but the rear had a different look. Because of the smaller turbine engine, designers were able to slope the rear decking backward, giving the operator a better rearward view. But the oil embargo and escalating fuel prices doomed the turbine 992 design by 1974, because turbine engines literally gulped fuel. Another design drawback was that the air going into the gas turbine had to be particularly void of dust particles to avoid damage to the engine. So fine were the filters utilized on the loader that they literally had to be changed hourly. Although refinements could have improved these limitations, the company did not feel it would be profitable to pursue turbine power.

The 992G carries a far larger bucket than the previous 992D model it replaces. With its 16-cubic-yard bucket, it is a full 2 cubic yards larger than its predecessor. It also has more power on hand, thanks to a new Cat 3508B EUI diesel engine, which produces 800 flywheel horsepower. *Urs Peyer*

The Caterpillar 992G went into full production in early 1997, and has proven to be just as tough as previous 992 series models. But with the extra features and power on hand, the 992G is from 9 to 22 percent more productive than the 992D, depending on application. Shown working side by side in September 1996 are a 992G and its bigger brother, the 994. *ECO*

As customers pursued greater productivity with off-highway hauler fleets of steadily increasing payload capacity, Cat engineers could see that their big loader would have to change to keep pace with the times. In late 1977, Caterpillar introduced its incredible 992C front-end loader. The 992C was far more than a dressed-up version of its predecessor. In truth, there was really no comparison between the old 992B and the new C-model. The 992C had faster cycle times, carried bigger payloads, and had more power and breakout force. It was not only Caterpillar's largest and finest loader to date, many in the world quarry and mining industries regard the 992C as one of the greatest large wheel loaders ever designed.

From any angle, the 992C looked like it meant business. Housed in the rear was a new V-12 Cat Model 3412, turbocharged and aftercooled diesel engine, rated at 690 flywheel horsepower (735 gross). Capacity had risen substantially to 12.5 cubic yards and 37,500 pounds. This was due in part to the redesigned lift arms, which included the use of Z-bar loader linkage design. Overall operating weight was listed at 198,260 pounds, when equipped with the standard rock bucket. The 992C could load a 50-ton-capacity truck in three passes, and an 85 tonner in five. It featured a "modular" design of its major components, which could be quickly removed for repair or replacement. An Electronic Monitoring System (EMS) was also standard, providing the operator with visual and audible alarms to problems in critical machine systems. Optional equipment included Caterpillar-designed steel shoe beadless tires for severe operating conditions, and a high-lift loader arm arrangement that allowed the 992C to production load 120-ton trucks. So well thought out was the design of the 992C that it went through most of its production life virtually unchanged. In 1982, the 992C's standard bucket capacity was increased to 13.5 cubic yards, but on the outside, it looked just like any other C machine. By the time the model was finally replaced by the redesigned 992D in 1992, over 2,500 C-model units had been put into service worldwide, making it the best-selling large mining loader of all time.

Introduced in 1993, the 990 model line filled the gap in capacity that was starting to develop between the 988F and 992D series of loaders. The 990 was a 610-flywheel-horsepower, 11-cubic-yard loader that weighed in at 161,994 pounds. The 990 looked much like the 992D, but was proportioned slightly smaller. The most notable difference was an updated ROPS cab. The original 990 was replaced in 1996 by a more powerful, 625-flywheel-horsepower 990 Series II–model offering. *Urs Peyer*

The 992D continued the success story of the 992C. Though the powertrain was virtually unchanged from the previous model, a host of improvements increased productivity measurably. The 992D now featured the Caterpillar-designed STIC (Steering and Transmission Integrated Control) System, which replaced the standard steering wheel and transmission shifting lever with a single joystick control mounted to the left of the operator. Using the joystick, the operator controlled all steering and transmission functions. The 992D also featured a Computerized Monitoring System (CMS) that kept an eye on all of the loader's key functions, a new contour air-suspension seat, and increased sound insulation in the ROPS cab. On the outside, the 992D was equipped with a larger 14-cubic-yard bucket, which allowed it to load a 95-ton-capacity hauler in four passes, and a 150-ton truck in six passes when equipped with the high-lift option. The loader also featured a newly designed operator's walkway over the left rear tire. In early 1996, the 992D got an increase in power from an improved Cat 3412C diesel engine, which was capable of 710 flywheel horsepower (755 gross). This would be the last change in this series before the introduction of an entirely new design by the end of 1996.

In September 1996 at the MINExpo in Las Vegas, Caterpillar introduced the radical 992G. Designed from the ground up as the replacement for the 992D, the 992G broke new technological design ground for the company. What made the 992G loader unique was its one-piece cast-steel box section front lift arm design, instead of the former twin-boom configuration. This new front end had three times the torsional strength of the previous model. Stresses were spread over the entire boom into the frame and not through weld joints. Though the mono-boom concept was not necessarily new to the industry, it had previously only been utilized on smaller machines, never on one the size of the 992G.

Along with the newly designed front end, the 992G had a host of other standout features: The structures on the loader are more than 90 percent robotically welded, for highly consistent welds with deep plate penetration and fusion resulting in increased strength. The new box section frame is extended further forward, improving rail-to-hitch strength, and giving the working platform more stability and balance. Mounted in the rear of the loader is a more powerful and cleaner burning eight-cylinder Cat 3508B EUI. This twin-turbocharged and aftercooled diesel engine is rated at 800 flywheel horsepower (880 gross). And like the last version of the previous model, the 992G came standard with the STIC System.

With all of these changes and extra power on hand, the capacity of the big loader also increased. The 992G's bucket capacity ranges from 15 to 16 cubic yards, depending on the specific application for the unit. Equipped with the 16-cubic-yard large standard spade-edge bucket, it was a perfectly sized machine for loading 100-ton-capacity haulers, such as the Caterpillar 777D, which it could fill in four quick passes. Equipped with the 16-cubic-yard bucket, the 992G weighed in at 202,499 pounds in full operating trim, with the high-lift version coming in at 209,343. By late 1999, these figures had risen to 207,519 and 215,823 pounds, respectively.

After the MINExpo trade show, preproduction 992G loaders were placed in operation around North America for testing purposes. In March 1997, Caterpillar officially released the new loader for sale to the world earthmoving marketplace.

Because of the 992 loader series' increased capacity over the years, a gap in the product line developed between the 988 and 992 model offerings. This was filled in 1993 with the 990 wheel loader series. The 990 looked much like the 992D, just proportioned smaller. But the 990 was not a small-scale performer. At the time of its introduction, the 990 was a 610-flywheel-horsepower (660 gross), 11-cubic-yard, 161,994-pound wheel loader. By 1996, power had increased to 625 flywheel horsepower (675 gross) with the use of the Cat 3412E T/A diesel engine in the 990 Series II. Compared to previous 992 model offerings, the 990 is quite a bit larger than the old 992B, but a little shy of the 992C's overall bulk. This extra model range helps customers choose a loader that is perfectly matched to their particular job requirements. If the 988 range is too small, and the 992 machines too big, chances are the 990 Series II will be just right.

The 994

As Caterpillar engineers were putting the final touches on the first 992 front-end loader, the company started work on an even larger machine identified as the 994. The experimental 994 was approximately twice the size of the 992, and was designed around a four-wheel-drive, articulated frame steering, diesel-electric drivetrain. First tested by Caterpillar in early 1969 at its Peoria Proving Ground, it was powered by the same engine found in the Cat-designed electric-drive trucks—the D348, V-12 diesel, rated at 960 flywheel horsepower (1,000 gross). Tires were 50x39-inch-series rubber jointly developed by Caterpillar and Goodyear. Bucket capacity was rated at 20 cubic yards and was of an ejector type, which was intended to reduce lift height. Bucket linkage was a unique four-bar design. But the troubles encountered in the off-highway electric-drive

The early Caterpillar 994 loader was a large and powerful machine. With its 1,250-flywheel-horsepower V-16 Cat 3516 diesel and standard 23-cubic-yard bucket, the 994 was one of the largest mechanical-drive front-end loaders available to the mining industry. Because of early tire wear problems, many customers installed tire chains on their 994 loaders to protect them from rock cuts and abrasions. *ECO*

truck program would cause the downfall of the electric-drive 994 as well. Both programs were canceled in late 1969. All work on the diesel-electric 994 officially ended in January 1970.

In 1976, product development tried once again to propose a large wheel loader, again using the 994 identification. Key features would include a modular design, a conventional mechanical drivetrain, and beadless tires only. The use of components found in other Cat designs was also a key consideration to reduce costs and improve profitability of the program. Early specifications on the proposal called for a Cat 3516 diesel engine, rated at 1,350 flywheel horsepower, connected to a four-speed, powershift transmission. The proposal also indicated a 21-cubic-yard ejector-bucket with a rated payload capacity of 63,000 pounds. Target weight of the new design was 279,000 pounds. The operator's cab would have been the same unit design utilized on the 988 and 992C loader programs of the day. Project personnel hoped to have the first experimental unit up and running by 1979, but Cat management had other ideas. Because of the costs involved, management felt the project was just too risky. The possible number of units sold would have been small to begin with, so any mistake would have meant big losses financially. The project was essentially killed

The largest front-end loader currently offered by Caterpillar is its model 994 series. The 994 was one of the first mining vehicle designs to be put into production by Caterpillar's MVC Engineering Division. In August 1990, the first prototype 994 was being built in its assembly bay at the Decatur plant. Though the prototype was assembled in Decatur, full production for the giant loader would be handled by Caterpillar's Joliet assembly plant. *Caterpillar, Inc.*

before it ever got started. In the end, only a large-scale model of the loader was ever built.

In the late 1980s, Caterpillar once again started development of a large wheel loader that was almost twice as large as the then-current 992C. The new design, again referred to as the 994, primarily targeted the large mining sector. Even though two previous concept designs developed by the company in the past were identified as 994, they had no bearing on this project. In 1989, production started on the first prototype 994 at Caterpillar's newly formed Mining Vehicle Center (MVC), located within the Decatur assembly plant. By September 1990, the completed loader was dedicated at the plant.

The 994 was the largest wheel loader the company had ever built. Equipped with a standard bucket rated at 23 cubic yards, it ranked with some of the largest wheel loaders ever produced. Weighing in at 385,297 pounds, its mass was similar to mechanical-drive loader giants of the past, such as the 389,500-pound Clark Michigan 675C and 396,720-pound Kawasaki/SMEC 180t. Although these two machines did carry slightly larger buckets than the 994 (24 and 25 cubic yards respectively), all took up about the same amount of space.

The 994 was designed to perform a variety of digging assignments at a mining site. It was built to be an aggressive, highly efficient primary loading tool, and in certain mining operations it was economically competitive with smaller cable and hydraulic shovels. The 994 could move quickly from face to face within a mining operation. This mobility also helped eliminate lag time between blasting, since the loader could work up to the last minute, and then return as soon as the blast was over and the all-clear was given. The 994 could also serve as a backup for a mine's large electric cable shovel fleet. If a shovel was down for repairs, the loader could be brought in right away to take up the production slack. Once the shovel was

ready to go back on line, the loader could quickly travel to another work site within the mine. The 994 loader's mobility is one of its strongest selling points. The 994 is designed to "pass-match" trucks in the 150-ton class and up. It can load 150-ton haulers in four passes, 195-ton units in five to six passes, and 240-ton trucks with seven to eight passes, when equipped with the high-lift loader arm arrangement.

Making all of this possible was a rear-mounted four-stroke Cat 3516 V-16 diesel engine, turbocharged and aftercooled, and capable of 1,250 flywheel horsepower (1,336 gross). This was mated to a three-speed powershift transmission. The all-wheel-drive system utilized a double-reduction, planetary final drive at each wheel, which reduced the torque stress placed on the axles. The loader was geared for a top speed of 13 miles per hour in top gear. The 994's standard 23-cubic-yard bucket was rated with a maximum 70,000-pound payload capacity. Other buckets are offered that are specially matched to a material's specific density, such as coal (45 cubic yards) and iron ore (18 cubic yards).

During its early months of operation, there was one area of the 994 that was clearly not up to the task, and that was the tires originally specified for the unit. The original 49.5-57 68 L-4 series mounted on 36-inch-wide rims were just barely adequate for a wheel loader of the 994's size and performance capabilities. But at the time, there just wasn't any other choice available from the tire manufacturers. The tires were basically reinforced hauler designs. Clearly, something else was needed. Soon, the tire manufacturers released a larger design, the 53.5/85-57 mounted on 44-inch-wide rims, built only for wheel loader use. The tire made an immediate difference in the loader's performance and productivity. During the first few years, other upgrades were introduced to improve the 994's performance and reliability, such as even larger 55/80 R57 radial

The 994 could be equipped with a number of custom-designed buckets to meet a customer's specific job requirements. This 994, operating in the Powder River Basin of Wyoming, is equipped with a giant 45-cubic-yard coal loading bucket. Introduced in 1995, this coal loading bucket was the first ever installed on a 994. *ECO*

Equipped with a standard 23-cubic-yard bucket, the 994 is able to load 150-ton-capacity haulers in four passes; 195-ton trucks in five to six; and 240-ton units in seven to eight. *ECO*

tires, improved torque converter, redesigned pump drives, beefed-up front axle mountings and lift arm assemblies, improved cooling capabilities, and greater bucket options. Through all of these improvements, the company retained the basic 994 model name.

More significant changes were in store for the big Cat loader in December 1998, when the company introduced an upgraded model in the form of the 994D. From the outside, little was changed on the loader, outside of the decals, the curved exhaust tips (replacing the flat top design), and extended air cleaner stacks. But internally there were significant changes. The new model was now equipped with the improved 3516B EUI diesel. This cleaner-burning diesel was still rated at 1,250 flywheel horsepower, but the gross rating was now up to 1,375. Other improvements in the 994D included strengthening key stress areas in the lift arms and frame, as well as a new air filtration system. Also, a new Rimpull Control System (RCS), which allows the operator to choose from four factory preset reduction rimpull settings, was made standard. The RCS helps extend transmission life and reduce costly tire wear. The steering controls were also revamped on the new loader. The steering wheel was replaced by a STIC System joystick control, mounted to the left of the operator's seat. Caterpillar also offers customers the chance to retrofit pre-D-series loaders with the STIC control system, along with other technical upgrade packages, including a new air-suspension contour seat. These changes and upgrades have made the 994D one of the most productive mining loaders in the industry today.

Improvements made to the 994 series over the years have steadily increased the loader's weight. Today's standard 994D tips the scales at 421,693 pounds in full operating trim, and 427,323 pounds when equipped with the high-lift option. When fitted with the optional 23-cubic-yard-wide spade edge bucket with MAA (Mechanically Attached Adapter) System and high-lift, the weight shoots up to 430,403 pounds, or just a bit over 215 tons. Make no mistake, the 994D is one mighty big loader.

Since its introduction in late 1990, the 994 series has gone on to become the best-selling large mining wheel loader in the world. With over 200 of the original models sold before being replaced by the D series, it holds a commanding sales lead over all of its competitors combined.

In late 1998, Caterpillar introduced an updated edition of its big mining loader, identified as the 994D. The 994D now featured the joystick-steering STIC System, which replaced the previous model's steering wheel. Pictured is a 994D chassis nearing the end of its assembly process. The 994D loaders were originally assembled at the Joliet plant, but production of the giant loader was eventually transferred to the Aurora, Illinois, facilities. *ECO*

Large Wheel Dozers

During the 1990s, Caterpillar made major strides in producing and marketing large mining equipment. These offerings were further bolstered in July 1997, when Caterpillar purchased the rights to the large rubber-tired wheel dozers produced by Tiger Engineering Pty, Ltd., of Australia. Tiger, with the assistance of Caterpillar Industrial Products, Inc., produced its first large, rubber-tired wheel dozer concept in 1980, at the request of Mount Newman Mining of Western Australia. Mount Newman was in the market for a large wheel dozer with the mobility of a Caterpillar 834 wheel dozer, with production capability equal to or greater than a D9-class track-type tractor. The first model to be designed by Tiger, identified as the 690A, was based largely on components from the Cat 992C wheel loader. These included the rear-end assembly, drivetrain, and cab. A new front-end assembly was fabricated by Tiger similar to the one used on Caterpillar's 834 wheel dozer, but on a much larger scale. After completion of the prototype unit, it was officially delivered to Mount Newman Mining in 1982. Three more would soon follow.

The early Tiger 690A wheel dozers incorporated the 992C's standard torque converter. It worked well enough, but Tiger engineers felt that another unit would be needed to expand the dozer's performance capabilities. In 1985, Tiger introduced the 690B wheel dozer with a torque converter sourced from the Cat 773B haul truck. The torque converter was modified with an electronically controlled lock-up clutch, which provided direct drive in all forward and reverse gears by "locking" together the torque converter's turbine and impeller. This arrangement increased tractive effort, produced higher travel speeds, and prevented the torque converter from overheating on long push distances.

The Tiger dozers worked mainly in open pit mining operations where they were used to clean

The 994D is powered by a four-stroke Cat 3516B V-16 turbocharged and aftercooled diesel engine, rated at 1,250 flywheel horsepower (1,375 gross) at 1,600 rpm. The engine is fitted at the factory and ships complete with the main rear section of the chassis. *ECO*

up the pit floor around loading shovels, and to knock down and level waste dumps. They were also utilized for maintaining large coal stockpiles at power generating plants. Because of its versatility, keeping the Tiger wheel dozer busy required a fairly large mining operation. By June 1992, only 22 Tiger dozers had been placed into operation, mainly in Australia and the United States, with one unit going to Europe. The Tigers were fantastic at what they were designed to do, but a bit too specialized for most of the mining operations of the day. But that was soon to change.

In 1993, the 690B became the 690D, after the Caterpillar 992D loader was introduced. The 690D now utilized a new Tiger front frame design that was based largely on front hydraulic components sourced from the D10N dozer. The drivetrain came from the 992D wheel loader. Of greater significance was the use of Caterpillar's joystick STIC System, which eliminated the need for a steering wheel. The 690D also came with Cat's CMS (Computerized Monitoring System) for tracking the dozer's drive-train and electrical systems. In the real world, operators simply loved the STIC control system, which

All vital operating and drivetrain systems are checked thoroughly at the factory before final disassembly and shipping begins. Though the operator's cab is in place on this unit, it will be removed before shipping because of height restrictions. *ECO*

The 994D looks like the previous model, except for the exhaust pipe tips and air cleaner housings. It is what you can't see that makes the improved loader more productive than its predecessor—joystick steering controls, a stronger frame, a cleaner-burning engine, new filtration systems, and an improved modular cooling system. The improvements have made the 994D loader a stellar performer in its size class. *ECO*

The 994/994D series is the best-selling ultralarge mining front-end loader in the world today. With all of the advancements and additions made to the loader since its unveiling in 1990, the model has tacked on a few pounds. The standard 994D weighs in at 421,600 pounds in full operating trim. A unit equipped with the 23-cubic-yard-wide bucket and high-lift front end tips the scales at 430,403 pounds. *ECO*

helped increase the Tiger's overall productivity dramatically. This was one of the many reasons the 690D became the best-selling Tiger wheel dozer to come out of Australia.

To help sales and widen the appeal of the big wheel dozer program, in 1994, Tiger released a smaller companion model called the 590. The Tiger 590 was based heavily on the Cat 990 wheel loader. It utilized the 990's drivetrain and rear end, along with the front hydraulic system from the D9N dozer, including the twin lift cylinders and the dual tilt/tip blade control system. While the 690B utilized a modified torque converter from the 773B program, the 590 made do with the unit found in the 990 loader, set up so that the impeller clutch was locked in the fully engaged position. STIC steering control was standard. Other than a slight decrease in power output to 590 flywheel horsepower, the engine module was the same as that found in the 990 loader. When the 990 became a "Series II" machine in 1996, the Tiger model graduated to the 590B series. All of the drivetrain improvements made to the 990 Series II were also incorporated into the 590B, including the Cat 3412E T/A HEUI diesel with its 625-flywheel-horsepower rating.

These new Tiger wheel dozers, incorporating the STIC steering/transmission control, really sold well. By 1994, over 75 units had been placed in the field worldwide. By 1997, this number doubled to over 150 Tigers in service. As mining operations increased in size, the economics of owning a Tiger wheel dozer, a machine that was both sold and serviced through the Cat worldwide dealer network, was a win-win situation for the customer. This success story was not missed at Caterpillar headquarters in Peoria. Soon, things were about to change for the wheel dozers from Australia.

In September, at the 1996 MINExpo, Tiger Engineering introduced an even larger wheel dozer than the 690D known as the 790G. The new model was based on the newly introduced Cat 992G wheel loader, supplemented with other front-end hydraulic components from the D11R dozer. This introduction surprised many at the show, since the 992G wheel loader was being displayed for the first time at the same show. The cooperation between Caterpillar and Tiger engineers made it possible for both machines to be introduced to the mining industry at the same time—a first for a Tiger dozer. The 790G looked so promising to Caterpillar that by July 1997, it made a deal with Tiger Engineering to purchase the rights to design and manufacture the 790G, as

well as the smaller 590B. By March 1998, both models were officially released for sale to the world mining marketplace carrying new identities. The former Tiger 790G was now known as the Caterpillar 854G, while the 590B became the 844. Both model lines were now 100 percent Caterpillar owned and controlled. Shortly after the purchase, fabrication of the new wheel dozers was transferred from Australia to Caterpillar's Aurora, Illinois, assembly plant. This plant is the home of all of the company's big wheel loader production, making the integration of the former Tiger machines into the manufacturing process almost seamless.

The Caterpillar 854G now sits on top of the wheel dozer product line as the company's largest and most powerful offering ever. Powered by the V-8 Cat 3508B EUI twin-turbocharged and aftercooled diesel engine found in the 992G loader, its 800 flywheel horsepower provides the muscle to tackle the largest dozing and stockpiling jobs—as long as they do not include sharp rocks, which are the Achilles' heel of a wheel dozer. Three principal blade options are offered for the 854G: For standard dozing assignments often encountered within a mining operation, semi-U-blades and heavy-duty semi-U-blades are available, both rated at 33.1 cubic yards, and 20 feet, 7-inches wide. For coal stockpiling work, a massive 58.2-cubic-yard blade, 23 feet, 7 1/2 inches wide, is available for high-volume dozing. Equipped with the heavy-duty blade option, the 854G weighs in at 220,424 pounds in full operating trim. When compared to the working weight of a standard D11R track-type tractor, which tips the scales at 230,100 pounds, one starts to get the picture of just how big and brawny the big Cat wheel dozer really is.

While the Caterpillar 854G was designed for use in the world's largest mining and coal stockpiling operations, the slightly smaller 844 model's main focus is smaller-scale mining facilities, along with general contracting and coal stockpiling work. Little changed in this wheel dozer class after the 590B became the 844. The 844's drivetrain and power output is the same as the 990 Series II wheel loader. And like the 854G, the 844 is offered with the choice of three dozing blades. The standard and heavy-duty semi-U-blades are both rated at 20.7 cubic yards, while the coal version carries a 40.2-cubic-yard rating. Widest of all the blades is the coal unit, measuring just over

The Tiger 690D, introduced in 1993 and based largely on the Caterpillar 992D loader, was by far the best-selling wheel dozer the company ever made. Having the same engine, power output, and STIC steering controls as its wheel loader cousin, the 690D soon became a "must have" in large open-pit mining operations the world over. *ECO*

19 feet in width. The heaviest operating weight version of the 844 is the unit with the heavy-duty semi-U-blade, which weighs in at 153,685 pounds. This is more than 42,000 pounds more than the third-largest wheel dozer in Caterpillar's product line, the 834B, similarly equipped.

The Competition

Caterpillar's large wheel loader and dozer programs primarily face competition from two outside sources today. For the large mining wheel loaders, Komatsu and LeTourneau are poised with very credible offerings that in some cases are quite a bit larger than the Cat machines. For wheel dozers, the only rubber-tired contenders are fielded by Komatsu.

Against Caterpillar's second-largest loader, the 992G, three machines, two from Komatsu, and one from LeTourneau, presently compete for market share. The LeTourneau offering, the L-1000, is a diesel-electric-drive, 925-gross-horsepower, 17-cubic-yard, 51,000-pound-capacity loader, with a

standard operating weight of 232,000 pounds. Introduced in 1982, it has enjoyed steady sales success, though not on the same scale as the 992 series. Sales have slowed recently on this unit, as mining operations seem to prefer the mechanical drivetrain layouts with bucket capacities in the 16- to 17-cubic-yard class.

Far more credible loader threats come from Komatsu in its WA800-3 and WA900-3 models. The mechanical-drivetrain WA800-series was introduced in 1986 and has been the closest rival to Caterpillar's 992 series, in power and capacity. The current version of the loader, introduced in 1998, is an 808-flywheel-horsepower, 14.4-cubic-yard, 222,955-pound machine. A high-lift version equipped with a 13.1-cubic-yard bucket is also offered. The WA800 loaders have always had a good reputation in the field and have sold in reasonably large numbers, but again, not on the scale of the Cat 992C/D/G model lines. But even so, the WA800-3 is a proven performer and defiantly keeps the engineers on the 992G loader project on their toes.

A slightly more powerful version of the WA800-3 is the WA900-3. Komatsu introduced this model line in 1996 as the WA900-1. Today's WA900-3 is an 853-flywheel-horsepower, 17-cubic-yard loader, with a listed operating weight of 226,810 pounds in standard form, and 227,760 pounds when equipped with the 15-cubic-yard high-lift front end. The WA900-3 utilizes the same basic chassis, drivetrain, and frame as the WA800-3, but has a more robust front end, higher-ply-rated tires, and more horsepower dialed in. The engine in both units is a Komatsu-built SAA12V140ZE-2, turbocharged and aftercooled diesel. Only the power outputs are different in the two model lines. The WA800-3 fills a gap in the marketplace left vacant in Caterpillar's loader product line between the 990 Series II and the 992G. One would image that the next-generation 990 will be upgraded in such a way that it would be able to take on the likes of the WA800-3. Only time will tell.

Caterpillar's current largest wheel loader, the 994D, faces extremely tough competition from both Komatsu and LeTourneau. Komatsu has one machine with slightly larger capacity than the 994D, and LeTourneau offers no less than four loader models that are larger than the big Cat.

Officially introduced to the world mining market in October 1999, the Komatsu WA1200-3 Mountain Mover is the largest wheel loader ever designed and built by the company. At one time, Komatsu, when it was a part of the Komatsu-Dresser joint venture company, did offer a 24-cubic-yard, mechanical-drive loader called the Haulpak 4000. But this wheel loader was essentially an updated International Hough 580 design, and was not created by Komatsu in the first place. The WA1200-3 is a complete Komatsu creation, through and through. Currently assembled in Japan, the WA1200-3 utilizes a mechanical-drivetrain layout, and is powered by a single Cummins QSK60, 16-cylinder, turbocharged and aftercooled

After Caterpillar purchased the rights to design and manufacture Tiger Engineering's big wheel dozers in 1997, the Tiger 790G, which was based on the 992G loader, became the 854G. Big and powerful, the 854G is one of the largest wheel dozers in full production available to the mining industry. *ECO*

With 800 flywheel horsepower at the rear, STIC steering control, and a 33.1-cubic-yard semi-U blade up front, the 854G is the ideal tool for land reclamation, stockpiling, and pit clean-up at a large mining operation. Weighing in at 220,424 pounds (with the heavy-duty blade option), the wheel dozer is the heaviest ever produced by Caterpillar. Originally assembled in Australia, all 854G dozers are now built in Aurora, Illinois. *ECO*

diesel engine, rated at 1,560 flywheel horsepower (1,715 gross). Standard bucket capacity is 26.2 cubic yards, and 23.5 when equipped with the high-lift option. Overall working weight of the standard machine when fitted with 65/65-57 tires is 463,400 pounds. The high-lift version weighs even more, tipping the scales at a whopping 470,200 pounds. Though it looks like the WA1200-3 has the 994D outgunned on all sides, one has to remember that the 994D's basic design has been around since late 1990. The Komatsu design is relatively new and takes advantage of all the latest technology in its design and manufacture.

Born and bred in Longview, Texas, LeTourneau, Inc.'s diesel-electric-drive wheel loaders have battled the 994 loader series from day one. One month after the first prototype 994 went into service, LeTourneau countered in November 1990, putting its own ultralarge wheel loader prototype into a mining operation. That machine, identified as the L-1400, was a 28-cubic-yard-capacity design, capable of an 84,000-pound bucket payload. Introduced with a power rating of 1,600 gross horsepower, today's model now boasts 1,800, with a working weight of 445,000 pounds (450,000 pounds for high-lift). Not long after the L-1400 went into service, LeTourneau released still another, larger loader model in December 1993, identified as the L-1800. This diesel-electric-drive machine is a 2,000-gross-horsepower, 33-cubic-yard, 100,000-pound-capacity loader, with a standard operating weight of 480,000 pounds (485,000 pounds for high-lift). Both of these entries in the ultralarge wheel loader mining market are substantially larger than the current Cat 994D.

In 1999, LeTourneau introduced a new generation of diesel-electric-drive loaders, starting off with a new model identified as the L-1350. The L-1350 is rated at 1,600 gross horsepower and carries a standard 26-cubic-yard bucket with an 80,000-pound load capacity. Operating weight is listed at 390,000 pounds. The loader is LeTourneau's first design to make use of digital system controls and joystick operation for steering, hoist, and bucket functions. This loader is close in size and capacity to the Cat 994D, as well as the WA1200-3 by Komatsu, which is a keen competitor with the Texas-based company.

But the wheel loader that LeTourneau introduced at the 2000 MINExpo is literally in a class by itself. That machine, the L-2350, is simply

the largest and most powerful front-end wheel loader ever designed and built. With 2,300 gross horsepower, and a standard rock bucket capacity of 53 cubic yards and 160,000 pounds, it is over twice as large as any non-LeTourneau sourced loader currently on the market. It features all of the latest digital controls introduced on the L-1350 series a year earlier. Weight for this machine is a staggering 540,000 pounds in full operating trim. It is probably a bit unfair to compare the 994D to the L-2350. Though both loaders are huge by industry standards, they are in different loading size classes and do not compete directly with each other.

Although these latest competitive wheel loader designs all seem to put a damper on the possible production capabilities of the 994D, one has to consider the age of the original loader design. Currently, Caterpillar engineers are in the advanced design stages of a new wheel loader of a significantly larger size that will someday complement the 994D. But at the time of this writing, most of the information is still top secret, with no announced release date on the horizon.

The competition to Caterpillar's largest wheel dozer design, the 854G, is relatively light when compared to the rubber-tired front-end loader program. In fact, it is limited to just one model—Komatsu's WD900-3 wheel dozer. The WD900-3 is based largely on the WA900-3 wheel loader, but utilizes a specially fabricated front end. While the WA800-3 and WA900-3 loaders destined for North American consumption are assembled in Peoria, Illinois, all of the WD900-3 wheel dozers are built entirely in Japan. Powered by the same 853-flywheel-horsepower Komatsu diesel found in its wheel loader counterpart, the dozer variation is offered with either a 34-cubic-yard semi-U-blade that is 21 feet, 3 inches wide, or a 24-foot, 3-inch-wide coal stockpiling unit that is capable of handling 58.9 cubic yards. Overall working weight of the WD900-3 is a healthy 220,460 pounds—right in line with the Cat 854G. The first WD900-3 wheel dozers went into limited service in 1999 at mining operations in Australia. The WD900-3 is strictly a special order item and is not even listed in the company's standard product guides. Currently, only a handful of these units have been delivered into service, and all are either working in Australia or the United States.

The very nature of a rubber-tired wheel dozer allows it to perform a multitude of tasks. Its mobility is one of its key selling points. As soon as one dozing job is completed, it can quickly move on to another location. What the 854G can not work in is very rocky material, which would cause severe tire wear and damage. In these conditions, the track-type tractor, such as the D11R, reigns supreme. *ECO*

Largest of all the Caterpillar mining excavators is its 5230 series. First introduced in mid-1994, the 5230 was originally released in a front-shovel configuration, with a standard bucket capacity of 22.2 cubic yards. Power output from its Cat 3516 V-16 turbocharged and aftercooled diesel engine was 1,470 flywheel horsepower. The 5230 FS can load this 195-ton-capacity 789C hauler in six quick passes. *ECO*

CHAPTER FIVE

The 5000-Series Hydraulic Excavators

5230 HYDRAULIC EXCAVATOR

During the 1960s, many of the industry's established earthmoving machinery designs and concepts were supplemented, and then eventually replaced, by what is now referred to as the hydraulic excavator. Before the advent of this machine, its role was performed in general contracting work, as well as in quarries and mines, by cable-type shovels or backhoes. These highly efficient and reliable designs had been the number one choice for most loading assignments. They were built by a large number of manufacturers over the decades, and most in the industry could not comprehend their eventual demise.

But the hydraulic excavator did not immediately win the hearts of the contractors at large. Many of the early machines were small and were plagued by leaky hydraulic systems and overwhelmed hydraulic pumps. To appease old and new customers alike, established builders of the time, such as Bucyrus-Erie, Insley, and Koehring, placed hydraulic designs alongside cable machines in their product lines.

Caterpillar's hydraulic excavator program all started with this full-size wooden mock-up 3/4-cubic-yard machine identified as the X10. Built in April 1969, it would lay the groundwork for the company's first production hydraulic excavator in 1972, the model 225. *Larry Clancy Collection*

As the 1960s came to a close, hydraulic excavators were approaching the size of many of the tried-and-true cable machines. In 1970, Poclain of France introduced a 10-cubic-yard hydraulic excavator known as the EC-1000. With its 150-ton operating weight, it signaled to the industry that the hydraulic concept was not only applicable to smaller and midsize machines, but larger mining shovels and backhoes as well. The demise of the contractor-sized cable excavator was at hand, leaving only the larger mining cable shovels to continue on.

During these changing times, Caterpillar didn't have to worry too much about updating old cable models since it had none to offer in the first place. But it was smart enough to realize that the hydraulic excavator was the future for this type of machine. Their speed and productivity were beyond question. And as time went on, their reliability records were also on the upswing. In the late 1960s, Caterpillar management approved development funds for the company's first hydraulic excavator program.

By early 1969, Caterpillar's Research and Design department had the first full-scale wooden mock-up, identified as the X10, built for management's approval. The first 3/4-cubic-yard pilot machine was completed in January 1970. Known by then as the 625X1, it would form the basis for the first production Caterpillar hydraulic excavator, the model 225. Officially introduced in 1972, the 225 was designed and built by Caterpillar

from the ground up, and not some type of hybrid, joint-effort machine tried by some manufacturers in the mid-1960s.

The model 225 was only the beginning. In 1973, Caterpillar introduced a slightly larger model in the form of the 235. This was followed in 1974 by the rather large 245 series, considered by many to be one of the company's best excavator designs of the 1970s and 1980s. The company rounded out its excavator model line with the smaller 215, built at Caterpillar's assembly plant in Belgium. To further broaden the appeal of its new excavators, the company offered the quarry and smaller mining industries front-shovel versions of the 245 in 1976, and the 235 in 1978. All of these models were well accepted in the industry, and they helped establish hydraulic excavators as one of the company's most important and profitable product lines for future sales.

As the 1970s came to a close, European manufacturers such as Demag, O&K, and Liebherr had a clear advantage when it came to the larger mining-size class of hydraulic excavators. For smaller construction and contractor-sized units, designs from Japan were quickly establishing themselves as the machines of choice. When the high value of the U.S. dollar made the business of exporting heavy equipment overseas very challenging, foreign competitors took full advantage. The high dollar also made it easier for overseas companies to move their goods into the United States. Companies such as Komatsu, Hitachi, and

The largest Caterpillar-designed hydraulic front shovel available during the 1970s and 1980s was the 245 FS. Introduced in 1976, it was based primarily on the standard 245 excavator model from 1974. Weighing in at 146,700 pounds, the 245 FS was available with a 5-cubic-yard front-dump bucket, or a 4-cubic-yard bottom-dump design. *Caterpillar, Inc.*

The prototype 5130 FS hydraulic mining excavator gets ready to be taken into the convention hall for the MINExpo '92 trade exposition in September 1992 in Las Vegas, Nevada. Though the 5130 was almost to production stage at this point, Caterpillar engineers made further changes to the design, rerouting hydraulic pressure lines on the main boom. *Caterpillar, Inc.*

Kobelco made inroads into what was becoming a very crowded U.S. marketplace for hydraulic excavators. Though Caterpillar had a strong line-up of machines, it was small when compared to the many offerings coming from Asian manufacturers. If Caterpillar wanted any hope of flourishing in this rapidly expanding marketplace, it needed to take drastic action.

In 1984, Caterpillar formed an alliance with Eder of Germany. Under the agreement, it would sell and service seven of the Eder compact hydraulic excavators in North America as Caterpillar products. Then, in 1987, Caterpillar started to import hydraulic excavators built by Mitsubishi of Japan. This was made possible by a joint working relationship between the two companies dating back to 1962, when Caterpillar and Mitsubishi Heavy Industries, Ltd., formed an equal-ownership manufacturing and marketing company called Caterpillar Mitsubishi Ltd. When the production of excavators commenced in Japan, the company's name was changed to Shin Caterpillar Mitsubishi Ltd. These new imported machines were sometimes referred to as the "E-machines," since their product nomenclatures all started with the letter E. The largest of these early designs was the E650 series. The E650, which was available as a front shovel or backhoe, was very similar in size to the model 245.

Early 1992 would see the launch of the new 300-series hydraulic excavators. These newly designed models now integrated the best in technology that Caterpillar engineers in the United States, and Shin Caterpillar Mitsubishi engineers in Japan, could produce. In 1993, the largest of these new machines, the model 375, was officially introduced as the replacement for the aging 245B Series II. Weighing in at approximately 178,800 pounds, it was considerably larger than either the 245 or E650 machines, which tipped the scales at 143,520 and 140,900 pounds, respectively. The 375 was rated at 428 flywheel horsepower, while the former 245B Series II was rated at 360, and the E650 at 375. The 375 turned

out to be a worthy successor to Caterpillar's former large construction and quarry-sized excavator lines and was considered one of the finest machines in its size class for sale in the world. But as good as the 375 was, Caterpillar saw room for improvement. In January 2002, it was replaced by the new model 385B series. Rated at 513 flywheel horsepower (552 gross), the 385B weighs in at 190,370 pounds in standard configuration, and 197,000 pounds in mass excavator (ME) form. The largest bucket option is the ME version equipped with a 7.75-cubic-yard unit.

Though Caterpillar produced world-class hydraulic excavators, most of the models introduced were for construction and quarry-type applications. This is not to say that these Cat excavators were not in use at mining operations around the world. They were, but not as the principal loading shovel. As large as the 245/375 series of excavators were, they were still far too small to act as a serious, high-production loading tool in a large quarry or open pit mining operation. Caterpillar realized that an entire new class of excavators would be needed to compete successfully against the likes of Demag, O&K, Liebherr, and Hitachi. These new designs would be built to withstand the type of tough working conditions often found in the mining industry, and would be true real-world mining front shovels and mass excavators. That new product line would be the 5000-series of hydraulic excavators.

The first of the new-generation Caterpillar mining excavators was the model 5130. Originally conceived at Caterpillar's MVC in Decatur, Illinois, along with the 793, 994, and 24H in the late 1980s, the 5130 was a totally new design. Instead of building an entire range of 5000-series models all at once, Caterpillar management and engineers decided to start with the market segment that could produce the largest sales potential, the 100-ton-capacity truck class. Since there were literally thousands of these trucks in service around the world at any given time, they would give an instant market to the 5130. In this respect, the

Caterpillar introduced a backhoe version of its 5130 mining excavator in late 1993. Referred to by Caterpillar as a "mass excavator," the 5130 ME can handle a standard bucket capacity of 13 cubic yards. An optional 17.8-cubic-yard coal loading bucket was also available. Maximum working reach was 48 feet, 9 inches. *ECO*

In 1997, Caterpillar replaced the original 5130 with a new "B" version. The 5130B featured a more powerful 800-flywheel-horsepower engine (755 for the 5130), and various upgrades and improvements to increase the excavator's overall reliability. Pictured in October 1997, this 5130B FS excavator was one of the first to be put into service. *ECO*

5130 was designed from day one to perfectly match up to the 777C (95-ton-capacity) truck, and its eventual replacement, the 777D (100-ton-capacity), with five-pass loading performance.

The 5130 design program and prototype build-up proceeded from 1989 to the machine's eventual world unveiling in September 1992 at the MINExpo, held in Las Vegas. The 5130 FS (Front Shovel) model was the first in the new model line to be put on display at the show. The 5130 was more than 2 1/2 times larger than a 245B Series II excavator. In its original design, the 5130 was equipped with an 11-cubic-yard shovel bucket, and was powered by a single Cat 3508 EUI eight-cylinder turbocharged and aftercooled diesel engine, rated at 700 flywheel horsepower. This engine could also be found in the 777C hauler and D11N dozer, which helped customers greatly when it came to stocking repair parts. Weight of the prototype unit was approximately 320,000 pounds. But after months of rigorous testing of the early field trial units, changes were made to the excavator in early 1994 that greatly increased its performance and overall productivity. The 3508 diesel had its power increased to 755 flywheel horsepower (815 gross). Capacity of the front shovel bucket went up to 13.75 cubic yards, and castings and high stress areas were beefed up, bringing the weight to 375,000 pounds. By the end of 1994, this figure had increased to 385,000 pounds, or almost 193 tons, for the front shovel.

Not long after the 5130 hit full-production status, Caterpillar introduced a smaller mining shovel called the 5080. Officially released in early 1994, the 5080 was a 6.8-cubic-yard front shovel based primarily on the successful 375 backhoe excavator range. In 2002, the 5080 was replaced by the 5090B, due in large part to the 375 series being upgraded into the new 385 model range in January of that year. *Urs Peyer*

CAT
5230
CAT

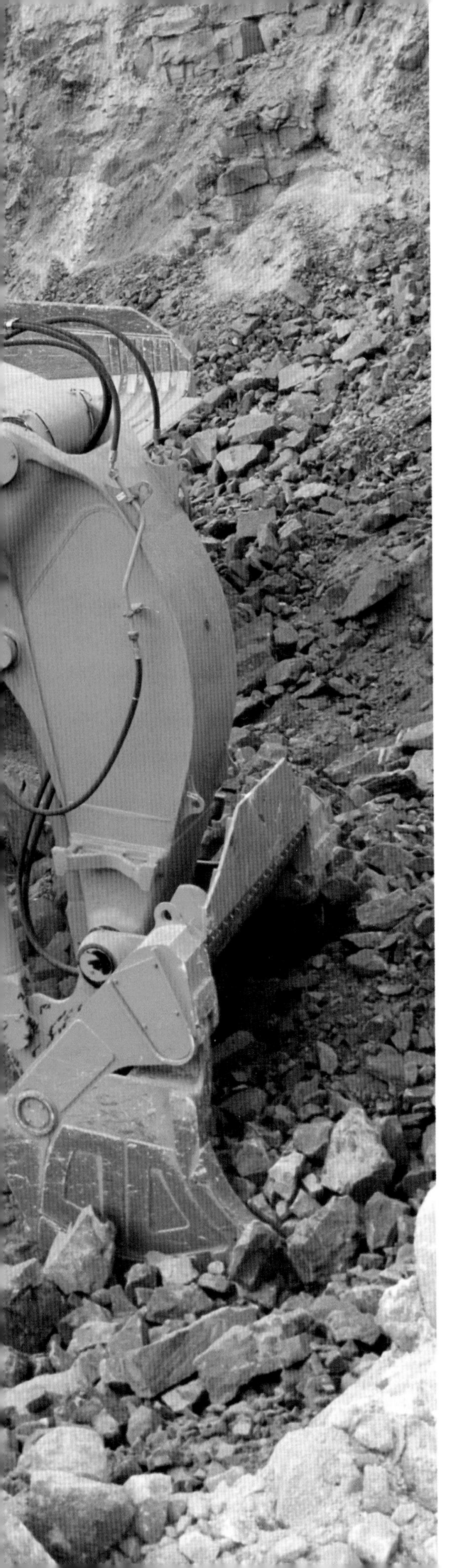

The company would not introduce a backhoe 5130 ME (mass excavator) configuration until late 1993. Equipped with either a 10.2-cubic-yard dirt bucket, or a 17.8-cubic-yard coal loading bucket, the 5130 ME proved to be a very productive and well-liked machine in the field—so much so that sales of the ME version soon eclipsed those of the FS model. The weight of the mass excavator was a bit more than the front shovel, with a fully operational unit coming in at 390,000 pounds.

The 5130 FS and ME versions soon became the best-selling hydraulic excavators in their respective size classes in the industry. It is true that a few of the earlier machines had design shortcomings, but with its worldwide dealer network, Caterpillar stood behind every machine until each performed to the customer's satisfaction. As new updates and improvements were made to later machines, these also would be incorporated into the earlier models, making them even more productive in the field. It is a tribute to Caterpillar engineers that the design of the 5130 matured as fast as it did. Many competitive manufacturers literally took decades to achieve what Cat had done in only five to six years.

Not long after the 5130 started to find its way in the marketplace, Caterpillar introduced its second 5000-series excavator, the 5080. Much smaller than the 5130, the 5080 was in fact the front-shovel version of the 375 backhoe excavator. The 5080 made use of the 375's hydraulic systems, as well as its power plant, the Cat 3406C ATAAC (Air-to-Air Aftercooled) turbocharged diesel engine, rated at 428 flywheel horsepower (455 gross). Though the main car body and undercarriage resembled those employed on the 375, they were suitably reinforced to withstand life as a front shovel in the 5080. Also of note, the operator's cab was raised substantially higher than on the 375, to give the operator a more unobstructed view during the loading cycle. Weighing in at 184,800 pounds equipped with a 6.8-cubic-yard bucket, the 5080 was perfectly matched to haul trucks in the 50-ton-capacity range. Because of its size, the 5080 was ideal for aggregate work in the European marketplace. In North America, customers seem to prefer a larger machine, such as the 5130, for this type of work.

In April 2002, the 5080 was replaced by the new 5090B front shovel, which is based heavily on the newly designed 385B excavator. Power output is the same as the 385B. Working weight is approximately 193,000 pounds when equipped with the standard 7.4-cubic-yard bucket. While

The 5230 FS was offered with a selection of bucket and Ground Engaging Tools (GET), and abrasion wear packages, that could be tailor-made to fit a customer's specific digging requirements. This 5230 FS is equipped with a rock bucket, best suited for working in shot rock, ore, and consolidated overburden. *Caterpillar, Inc.*

The 5230 FS utilized some of the largest steel castings ever fabricated by Caterpillar for one of its equipment designs. Weighing in at 702,000 pounds, the 5230 FS was certainly a machine with some presence. *ECO*

The working range of the 5230 FS was impressive to say the least. It had a maximum digging height of 48 feet, 5 inches, with a maximum reach at ground level of 46 feet, 2 inches. Maximum loading height was 33 feet, 8 inches. *ECO*

the original 5080 was assembled at Caterpillar's Aurora plant in Illinois and at its facilities in Belgium, the new 5090B will be built strictly in Belgium for the time being.

After the launch of the original 5080, Caterpillar introduced its massive model 5230 mining excavator series. The 5130 was a fairly large machine, but the 5230 was almost twice its size. The 5230 was designed to perfectly match up with the 150-ton-capacity Cat 785B, which it could load in four to five passes, depending on its configuration. But the big excavator was also equally at home loading the Cat 789B in five to six passes, and the 793B in seven to eight passes. Officially launched in August 1994, the 5230 FS front shovel was one of the largest pieces of earthmoving equipment ever designed by Caterpillar, incorporating some of the largest steel castings produced by the company up until that time. The 5230 was powered by a single Cat 3516 EUI quad-turbocharged and aftercooled diesel engine, capable of producing 1,470 flywheel horsepower

It wasn't long after the introduction of the original 5230 FS that a mass excavator version would soon join the line-up. Released in 1995, the 5230 ME originally was offered with a standard general purpose bucket rated at 20.3 cubic yards, with the option of a 31.4-cubic-yard coal loading version. By 1997, these capacities had increased to 23.5 and 36 cubic yards. *Caterpillar, Inc.*

(1,575 gross). This same engine could also be found in the Cat 789 and 793 series trucks, as well as the 994 wheel loader. Capacity for the front shovel model was 22.2 cubic yards, while the backhoe or mass excavator (ME) version, which was released in 1995, utilized a 20.3-cubic-yard general purpose bucket, with an optional 31.4-cubic-yard coal loading version. Overall working weight of the original 5230 model line was 693,800 pounds for the FS version, and 692,320 for the ME configuration. By 1997, these weights had increased to 702,000 and 697,980 pounds for each model line. The 5230 came standard with Automatic Engine Speed Control (AESC), which reduced fuel consumption by lowering the engine speed from 1,750 to 1,350 rpm if the hydraulic controls were not actuated for 4 seconds. The operator's cab also came equipped with the Vital Information Management System (VIMS), which continuously displayed critical machine performance data. For shipping, the 5230 breaks down into 11 basic modules. Originally assembled at the Joliet plant in Illinois, the 5230, along with the 5130 excavators, is now the responsibility of the Decatur plant, which is the home of the company's mining truck operations.

Once the 5230 program was well under way, Caterpillar engineers set their sights on a new model program for an excavator positioned just below the 5130 in size, identified as the 5110. The 5110 was designed to team up with haul trucks in the 45- to 70-ton-capacity range, such as Caterpillar's own 773 and 775 series models. Power would come from a single Cat 3412E twin-turbocharged and aftercooled diesel engine, rated at 600 flywheel horsepower (640 gross). Maximum bucket capacities would be set at 9.5 cubic yards for the backhoe ME configuration, and 9.8 for the front-shovel FS version. Overall working weights would be 276,728 pounds for the ME and 271,215 for the FS. Final specifications were set on the 5110 program in late 1995, with the first prototype unit, a backhoe model, completed at Joliet in early 1996. Initially only eight of these prototype field-follow machines were built, seven backhoe ME configurations and one front-shovel model. All were thoroughly tested in large construction, quarry, and mining sites across North America for most of 1996 and 1997 to obtain real-world field performance data. Following the testing program, all eight units were brought back to the factory or proving grounds for complete tear-down and inspection of every single component. The market segment that the

The 5230 ME has proven to be just as popular in the mining industry as the front-shovel model version. As far as sales go for each model, it is currently a dead heat between the two types in the marketplace. This 5230 ME working in July 1997 in Pennsylvania, can load this 150-ton-capacity Cat 785B hauler in only four passes. Maximum reach for the ME configuration is 58 feet. *Urs Peyer*

5110 would be competing in was an extremely competitive one, with many offerings from other manufacturers. Caterpillar knew that the machine that would finally be released would not only have to match the competition, but surpass it. In short, these eight preproduction machines sacrificed themselves in order to make the full production model the best it could be. That machine would be the 5110B.

The final production 5110B model line was officially launched at the MINExpo show held in Las Vegas in October 2000. Though the gestation period for this model line was a bit long, the final product was well worth the wait. Lessons learned from the pilot machines led Caterpillar engineers to make a host of changes that substantially increased performance and efficiency. The new 5110B is now matched to haulers in the 50- to 60-ton range, most notably the Cat 773D and 775D, but it has the muscle to effectively load trucks as large as the 100-ton-capacity 777D. The 5110B is powered by a Cat 3412E HEUI turbocharged and aftercooled diesel engine, rated at 696 flywheel horsepower (758 gross), substantially more than the pilot machine's output. Standard bucket capacity is up to 9.9 cubic yards; weight, with a fully operational backhoe, is 275,000 pounds. At this time, only a backhoe configuration is available. The company plans to release a front-shovel version in the near future, but no firm release date has been set, due to current market conditions at the time of this writing.

In July 1997, Caterpillar introduced an improved version of its original 5000-series excavator identified as the 5130B. More power was now available from a revamped 3508B EUI diesel, with an increased output of 800 flywheel horsepower (860 gross). Capacity of the front shovel was now rated at 14.5 cubic yards, with the backhoe

In late 1995, Caterpillar announced that it was working on a new excavator model range to be called the 5110, which would fit in the product line between the 5080 and 5130. By early 1996, the first of eight preproduction machines had been built. In all, seven backhoes and one front-shovel model were tested. The data gathered from these test machines would eventually lead to the full-production unit, the 5110B. This is the only 5110 FS built, pictured in 1997. Bucket capacity was 9.8 cubic yards, with a working weight of 271,215 pounds. After its field trials were through, it was dismantled and shipped back to Caterpillar for evaluation. *Urs Peyer*

version available with multiple bucket choices, ranging from a 13.7-cubic-yard rock bucket to a big 24-cubic-yard coal model. Key improvements included beefed-up swing bearings, new hydraulics and valving layouts, and rehinged ME buckets and cylinders, which increased the breakout force by 10 percent over the original design. As the modules of the excavator kept getting stronger, operating weight went up accordingly, with the ME version coming in at 401,000 pounds and the FS model tipping the scales at 399,000 pounds. The backhoe 5130B ME was the first version to be built by Caterpillar and put into service. The first front shovel started working in October 1997.

It would not be until November 2001 that the big 5230 would be released in an updated "B" model version. The first 5230B was officially unveiled to the industry at a special press preview held at the Decatur plant on November 13. And with a machine the size of the 5230B, there was a lot to see. Though the new "B" model looked much like the previous version, there were a host of improvements and upgrades to enhance the excavator's overall productivity and reliability. Power is supplied by an improved, clean-burning Cat 3516B EUI turbocharged and aftercooled V-16 diesel engine, the same as that found in the 789C, 793C, and 994D model ranges. Power ratings for this engine, as installed in the 5230B, are 1,550

CAT
CATERPILLAR
773D
CAT
773D

flywheel horsepower (1,652 gross) at 1,800 rpm. Bucket capacity is listed at 22.2 cubic yards for the front shovel, and 21 cubic yards for the mass excavator. Payload capacity is rated at 34 tons. Other upgrades include new cylinder castings; improved final drives, car body, and undercarriage; new swing drives and swing bearings; strengthened operator cab; and new hydraulic pumps and hydraulic line routings. For extra operator safety, a new ladder design is mounted just to the left of the cab. The 5230B also incorporates Caterpillar's exclusive Proportional Priority Pressure Compensating (PPPC) valves, which prioritize hydraulic fluid flow based on the operator's joystick inputs. The valves deliver only as much flow as the operator asks for, which saves fuel and reduces heat generation within the closed system. With all of these improvements comes a little extra weight. Approximate working weight of the 5230B FS is 721,000 pounds, with the ME version coming in at 723,400 pounds. The first 5230B built was a front-shovel FS model. The second unit assembled was a mass excavator (ME) version.

All of the larger 5000-series mining excavators are assembled at Caterpillar's Decatur assembly plant. This includes the main housing, car body, swing circle, and engine module. Components such as buckets, main boom and stick, crawler assemblies, and counterweights all ship from their point of origin at different Cat plants directly to the customer. The 5230B, 5130B, and 5110B are only then assembled into complete machines. The exception to this was the prototype 5230B, which was completely assembled at the Decatur plant for its world unveiling.

In a very short time, the 5130 and 5230 series excavators have gained the dominant sales position in their respective size classes in the mining industry. Though the 5110B has only been out since 2000, it is already making strong inroads into a very competitive class of machines, and is currently on track to leave its mark on the industry as well. Currently, three 5130B ME models are sold for every FS version built. As for the 5230 series, the ratio of ME models to FS configurations is a dead heat. As for the 5110B, only standard backhoe and ME versions are being manufactured.

But what of the future for Caterpillar's large mining excavator program? In 1995, the company was in the early design phase of what was referred to as the 54XX project. This project, which came to light in 1995, was a 30-cubic-yard-class machine, which would have targeted the 200- to 240-ton and larger capacity class of

Officially introduced at the October 2000 MINExpo, the 5110B was ready for worldwide release. The 5110B is powered by a single Cat 3412E HEUI V-12 diesel engine, rated at 696 flywheel horsepower (758 gross). Standard rock bucket capacity for the backhoe version is 9.9 cubic yards, with a 275,000-pound operating weight. The 5110B was designed to be matched with haulers in the 50- to 60-ton class, such as this 58.4-ton-capacity Cat 773D. *Caterpillar, Inc.*

The 5110B only requires four passes to fill this 773D hauler. It is currently available in standard backhoe configuration, or as a mass excavator, whose standard bucket capacity is 11.2 cubic yards. A front-shovel configuration has been planned for release sometime in the near future, but at the present time, no definite release date has been set, due to uncertain market conditions. *ECO*

It is late October 2001, as the first 5230B goes through its final engine tests at the Decatur assembly plant. In a couple of weeks, it will be ready for its world unveiling. In shipping, the 5230B breaks down into individual modular sections for easy transport. The main superstructure, cab, and swing circle all ship from the Decatur facilities. The crawler assemblies, bucket, and main boom all ship to the customer's assembly site from other Cat manufacturing plants. *ECO*

haul trucks. The 54XX project was to have taken dead aim at the Komatsu-Demag H455S and the O&K RH200. But the market for this size never materialized in the mid- to late-1990s. With the enormous costs involved in designing a machine of this size, and the lack of interest by the mining industry in purchasing any in numbers, Caterpillar quietly put the 54XX program on the back burner. The company has no current plans to build anything larger than the 5230B unless the marketplace for this type of loading tool increases dramatically.

The Competition

There is one word that best describes the type of competition that Caterpillar's larger mining excavators face in the earthmoving industry: fierce! Even though the company has made major strides into the large quarry and mining markets with its 5000-series machines, the competition likewise has been upgrading and introducing new models at a record pace. Companies from Europe, and especially Japan, produce some formidable challengers to Cat's big mining excavator lines.

The 5110B faces assaults from Liebherr and its proven R984B. Hitachi has recently introduced its new EX1200 excavator, which clearly has the 5110B in its sights. And Komatsu offers up its PC1100-6. All of these machines offer similar

The 5230B is the largest and most advanced mining excavator Caterpillar has ever produced. With a host of upgrades, such as improved final drives, car body, and undercarriage, to new cylinder castings, swing drives, and hydraulic pumps, the 5230B promises to be not only the most productive mining machine in its class, but also the most reliable as well. The future definitely looks bright for this big "CAT." *ECO*

On November 13, the first 5230B was announced at a special event held for the world heavy-equipment trade press. The 5230B is powered by a cleaner-burning and more powerful Cat 3516B EUI V-16 turbocharged and aftercooled diesel engine, rated at 1,550 flywheel horsepower (1,652 gross). The bucket capacities are virtually the same as the previous model, but overall working weight is now up to 721,000 pounds for the front shovel, and 723,400 pounds for the mass excavator version. *ECO*

working and performance characteristics to that of the 5110B. Since the 5110B has only been in full production since 2000, only time will tell who will be the victor in this size class.

The 5130B faces as much competition in the marketplace as the 5110B faces. But over time, the 5130 series has become the number one seller in its respective size class. Its major competition comes mainly from the likes of Hitachi and Komatsu. Hitachi has countered the Cat excavator over the years with its EX1800 series of mining shovels and backhoes. Recently, the model range was upgraded into the EX1900, with more power and slightly larger bucket options. The Komatsu PC1800-6 is also a relatively new design, which also offers performance on par with the 5130B. Other contenders include the Terex/O&K RH120E, the Liebherr R994, the Hitachi EX2500, and the Komatsu PC3000. All of these machines are larger than the 5130B, but smaller than the 5230B, and can compete against both designs. It really comes down to the customer's operation, and the size of its haul fleet.

On the large end of Caterpillar's mining excavator product line, the 5230B faces few direct competitors, but that does not mean they are any less formidable. The 5230B's main concern is Hitachi's EX3500/3600 family of "Giant" excavators. The latest model in this line, the EX3600, was announced in late 2000, with the first machines delivered in 2001. Weighing in at a hefty 772,000 pounds, the big Hitachi outweighs the 5230B by 51,000 pounds in front-shovel form. Standard bucket size is 27.5 cubic yards. The 5230B's is 22.2 cubic yards. Which machine a particular customer selects really depends on the final job application and the type of hauler fleets the excavators would ultimately be serving. Though far larger than the 5230B, the Komatsu PC4000 can also be a threat to the big Cat. But for the most part, they rarely compete head-to-head in a bidding war.

Ultralarge hydraulic excavators, such as the Terex/O&K RH400, Komatsu PC8000, Hitachi EX5500, and Liebherr 996 are too large for the likes of the 5230B and are in a class all by themselves. These machines compete with large cable mining shovels and are often the primary loading tool within major open-pit mining operations. They also sell in very small numbers. The 5230B simply cannot compete with the likes of these giants, but then again it was never designed to. If the market ever improves for this class of mining excavator to a point where it is profitable for Caterpillar to compete in it, you can be sure that a 5000-series machine of some substantial size will be there slugging it out with the best of them.

INDEX